Is Jesus Reasonable?

A Logical Analysis of the Statements Attributed to Jesus Christ

Gerald R. Prichard

UNIVERSITY PRESS OF AMERICA,® INC.
Lanham • Boulder • New York • Toronto • Plymouth, UK

University Press of America,® Inc.
4501 Forbes Boulevard
Suite 200
Lanham, Maryland 20706
UPA Acquisitions Department (301) 459-3366

Estover Road
Plymouth PL6 7PY
United Kingdom

Printed in the United States of America
British Library Cataloging in Publication Information Available

Library of Congress Control Number: 2009928990
ISBN: 978-0-7618-4650-5 (clothbound : alk. paper)
ISBN: 978-0-7618-4651-2 (paperback : alk. paper)
eISBN: 978-0-7618-4652-9

To my wife, without whose encouragement and continued support this book would have remained but a dream;
To my parents, whose Biblical example and teaching lit a lifelong pursuit of knowledge and a love for the Word of God; and
To my sister, whose daily courage and faith is an inspiration to trust in the unseen.

Contents

Preface

In writing the first four (4) chapters of this book (that is, aside from the applications in the fifth and final chapter), my technical position as a researcher has been to try as best as possible to take what I will call a neutral stance with respect to the "statements attributed to Jesus Christ" and with respect to formal human reasoning. Moreover, I think no comparative study of this specific nature could be executed with integrity by any other means. My method for maintaining such neutrality has been to: a) make no assumptions *a priori* regarding the 'validity' of the statements attributed to Christ from the perspective of formal human reasoning; and b) *vice-versa*, to make no assumptions *a priori* regarding the 'validity' of the formal human reasoning relative to the statements accorded to Christ. Notably, formal human reasoning consists solely of the set of axioms and resulting logical forms that bind all of scientific inquiry, and is significantly different from the logic employed in actual human's verbal statements, and especially from what would widely be known as human's "common sense"—both of which often contradict formal, human reasoning.

I assumed this neutral stance in order to help insure that a rigorous analysis could be documented, particularly given that the basis for any logical comparison cannot assume what it intends to demonstrate, either in part or in the whole. I think it only right to state that my own personal views of the Holy Bible, including its 'red-letter' statements that are accorded to Jesus Christ, are that of unequivocal support for inerrancy, infallibility, and immutability of the Holy Bible. However, for the purposes of this analysis, I have attempted as best I as humanly possible to put aside my personal views. Further, I did not know beforehand that a very strong majority of the statements attributed to Christ in the Holy Bible would be so consistent with formal human reasoning nor that its differences would be so significantly specific to the matter of double negations not always holding.

My reasons for writing this book are:

1) I sensed in my spirit the Lord's Calling to document what began as research into the generally conditional nature of the promises in the Holy Bible. This led me to investigate the promises attributed specifically to Christ, though I definitely only see this as a beginning which would not be completed until all of the conditional promises of God the Father in the Holy Bible were also examined with a rigor similar to that used for Christ's statements.

2) Beyond documenting the conditional nature of the promises accorded to Christ, I think that the Lord also wanted me to use my training in abstract mathematics and in cognitive psychology to:

a) demonstrate in a highly formal manner why the use of the strict rules of human reasoning (not just human thought or 'common sense') alone will fail to explain all of the conditional promises of Christ and as such be accurately applied to the sixty (60) conditional/procedural statements of Christ where differences with formal human reasoning exist; and

b) provide a comprehensive analysis of the most fundamental human conditions required by Christ's Statements in the Holy Bible (that is, those which if met would result in all of the promises of Christ being realized in the life of a Christian believer).

Ultimately, my view is that the findings of this book demonstrate that there is very significant and patterned consistency between formal, human reasoning and Christ's Words, but that where there are 'exceptions' and 'violations' I think of this as demonstrating a logical view of "science plus faith", not "science versus faith". I am confident that some of my fellow conservatives in the Christian community will think that, especially the 'violations' are intended to point out contradictions. I think of noting such differences as intellectual honesty which, rather than pointing to weaknesses in the statements attributed to Christ, actually points to the limitations of formal human reasoning. Gödel's Incompleteness Theorem is strong technical ground to support this view as his theorem, which is not a theory but is rather a law of formal logic, states essentially that any set of formal axioms (devised by human reasoning) will always have unanswerable questions.

Thus, from my perspective the 3% of the Christ's Words which yield either 'exceptions' or 'violations' compared to formal human reasoning are actually pointing to questions that formal reasoning simply cannot address in that formal human reasoning is axiomatic by nature. It is certain that those who see formal, human reasoning as comprehensive in its scope with respect to all logical inferencing will not agree with my perspective or interpretation of these results, but I would submit that the results themselves, minus any strictly logical interpretation of these results, will be found to technically sound within the scope of the statements attributed to Christ, though no objective analysis is known at this time to support a more generalized statement of this type when considering the rest of the Holy Bible as such has not been studied or at least not documented to date, to the best knowledge of this author.

Also, as summarized in my second reason for writing this book, I think that a key implication of these findings is that any theological hermeneutic which attempts to employ in a general manner or even more so is based upon formal, human reasoning will fail when applied to the 3% of the statements attributed to Christ containing either type of difference (that is, 'exceptions' or 'violations'). This does not mean that selective use of formal, human reasoning cannot ever be applied to theological interpretations or implications of the statements accorded to Christ, but it does mean that care must be taken to insure that such uses do not involve the specific differences documented in detail here.

Finally, in regard to the fifth chapter of this book, I want to state here, as I do in that chapter as well, that I have not remained neutral there in regard to the theological implications of the statements attributed to Christ, though I do think that even in Chapter 5 there is no logical implication made that cannot be supported by the analysis in the preceding chapters. As such, I view Chapter 5 as an application of the preceding chapters, and the "questions and answers" will hopefully be a help to those seeking to understand the foundational human attributes or actions required to realize the conclusions/promises attributed to Christ in the Holy Bible. For Christians my sincere hope is that they will find this to be not another list of human rules to obey, but rather as a help in determining how they can more effectively realize the blessings of living according to the teachings of Christ. For all others my hope is that the last chapter will serve as a further evidence of the highly structured and consistent logic employed in the statements accorded to Christ.

Chapter One

Overview

The foremost purpose of this study is to document and compare the "if, then" statements (both direct and inferred) attributed to Jesus Christ in the Holy Bible with each other and with the formal rules of human logic. More generally, this book is intended as a means of answering the question of whether Jesus Christ's statements in the Holy Bible can be considered reasonable based on a strict comparison using only literal meanings for words attributed to Christ in the Holy Bible to all of the formal rules of human reasoning. Having completed these technical documentation and comparisons through the first four (4) chapters, the second purpose is to specify the set of conditions (or 'axioms') that are not implied by any other condition(s) in the words attributed to Christ.

The answer to the foremost question fits the general pattern of the few 'Exceptions' noted. In particular, the answer to the question: "Is Jesus Reasonable" is mainly yes, but sometimes no.

The answer of 'mainly yes' applies to both the degree of logical consistency within the words attributed to Christ in the Holy Bible and also when compared to the formal rules of human reasoning. The answer of 'no' to the question of whether Jesus Christ can be called 'reasonable' also applies to both the internal and external comparisons. Specifically, thirteen (13) 'exceptions' are found internal to the words attributed to Christ in the Holy Bible, and fifty-one (51) 'violations' found by comparing the words attributed to Christ in the Holy Bible to all of the rules of formal human reasoning.

Somewhat more specifically, the primary finding of this comparative research is that, while the words attributed to Jesus Christ cannot be completely described as either wholly reasonable or wholly unreasonable in the strictest sense of human logic, these words and associated statements are generally consistent with regard to the rules of human reasoning. Indeed, there are only ten (10) sets of conditional/procedural statements attributed to Christ in the

Holy Bible that do not match the formal rules of human reasoning, which leads to the second most consistent finding from this analysis. Specifically, the eleven (11) sets of statements attributed to Christ in the Holy Bible that do not follow the formal rules of human reasoning all result from one (1) common logical issue (namely, that 'Not A' and 'A' can have the same 'truth value') and from three (4) theological issues (specifically, 'Trinitarian issues' most commonly; next an issue specific to the nature of Christ's Blood; 'issues of free will and Sovereign election'; and a question of the relation between "know(ing) God the Father and Christ" and "(God the Father's) commandments").

THE FOUR (4) TYPES OF 'EXCEPTIONS' EXCEPTIONS IN CHRIST'S USE OF LOGIC TO STRICT RULES OF HUMAN REASONING

1) The negation of the negation of a word or phrase is not equivalent to the original word or phrase (that is, {NOT (NOT A) is not necessarily equal to A}, which is logically equivalent to both 'A' and 'NOT A' being true or false at the same time. This type of exception is noted in nine (9) statements attributed to Christ.

a) First, Luke 6:45, Christ said that ". . . a good man out of the good treasure of his heart bringeth forth that which is good . . ." (which presupposes that there is a "good" man since the statement is not assumed to be vacuously true only), but then Matthew 19:17 says that "none is good, save one, that is, God"). Strictly speaking by the strict rules of human reasoning, both statements cannot be true simultaneously, and so this results in a exception according to the rules of formal human reasoning. See # 649 and # 909 in Appendix A. (Note: The same original Greek word for "good" is used in both statements).
b) Second, in this same vein, see John 1:18 compared to John 14:7-9, where "no . . . man hath seen the Father" and a set of persons are said to have "seen him (the Father)". See # 1095 and # 1249 in Appendix A.
c) Third, see John 14:28 (Christ says, ". . . my Father is greater than I.") compared to John 10:30 ("I and my Father are one."). See # 1271 and # 1412 in Appendix A.
d) Fourth, there are also two (2) instances in which Christ says, "If one receives or believes on me, then that one does not receive or believe on me, but rather on him who sent me". This statement is only true vacuously according to strict rules of human reasoning since the logical form (IF A, THEN NOT A) can only be true as a statement in formal human reasoning if 'A' is false or if both A and NOT A are true or false at the same time. See # 1023 and # 1210 in Appendix A.

e) As essentially another instance of the same type as 1.d), see John 7: 15 for an 'Exception' in which Christ is attributed as saying, "My doctrine is not mine, but his that sent me". (See #1475 in Appendix A.)

f) In John 11: 4 and 14, where Christ is to have said, "This sickness (for Lazarus) is not unto death" in verse 4, but then in verse 14, "Lazarus is dead". Using the main assumption of this study, which is literal interpretations only, this is another instance of multiple 'truth values' for Lazarus' death. A key observation though is that all other major translations than the one used in this study, which employed the King James Version, state effectively in verse 4 that "This sickness is not to end in death", which removes the violations of formal human reasoning here given that Lazarus is assumed in the words attributed to Christ in John 11: 44 to be alive again in saying, "Loose him (Lazarus), and let him go", which is after verse 14's pronouncement of Lazarus' death. (See # 1485 and # 1486 in Appendix A.)

g) In Mark 9: 37 (see # 480 in Appendix A) and in John 12: 44 (see # 1210), persons can both "receive" or "believe on" Christ and also not believe on Christ.

h) From the statement, ". . . nothing shall by any means hurt you." in Luke 10: 19 is in conflict with Matthew 24: 9 in which the disciples are told that when they are "delivered up to be afflicted . . . (they) shall (be) kill(ed) . . ." (See #1627 and #290 in Appendix A).

i) There is a quite similar 'exception' to 1.h) found in Luke 21: 16, in which ". . . some of you shall they (parents, brethren, kinsfolks, and friends) cause to be put to death" that is realized since Luke 21: 18 also states that, "But there shall not an hair of your head perish". (See # 1628 and # 1629 from Appendix A).

(*)—It is vitally important to note that the words, 'exception' and 'exception', is being used as opposed to the term, 'violation' since violations are only appropriately employed when comparing statements that are based on formal rules of human reasoning, not a mixture of potentially different types of logical axioms or rules.

2) The contrapositive of a logical form in Christ's Words is not always true when compared to strict rules of human reasoning, but this does not imply that the contrapositive is even generally found to be typically false when comparing the words attributed to Christ's Words to each other. {KEY NOTE: There is one (1) example of where the contrapositive of one of Christ's statements (when inferred using an "if, then" form) is explicitly stated in another of the statements attributed to Christ as being false. This exception is found in comparing John 4: 48 to John 20: 29, though there are several others where the strict rules of formal, human logic would indicate that the contrapositive does not always hold in the statements attributed to

Christ.} Such an 'Exception' is effectively an application of Exception 1) since the only means in formal reasoning by which a conditional statement and its contrapositive can have different 'truth values' is for a specific instance to permit both 'NOT A' and 'A' to be true simultaneously, which is the underlying cause of Exception 1). See # 1045 and # 1396 in Appendix A.

a) Special note: Of the 315 forms involving "GAMR" (in which the condition(s) are attributed to God the Father, Jesus, or the Holy Spirit and the conclusion(s) are attributed to humans only), the contrapositive of 100 (32%) of these forms is not considered to be false when compared to strict rules of human reasoning.

3) The converse of a conditional statement formed from an equivalence relationship is not always true. The only example of this type that can be demonstrated using solely the words attributed to Christ is found by comparing John 3: 6b, John 4: 24, and John 17: 3 (see # 1456, # 1458, # 1459 in Appendix A, respectively). According to the strict rules of human reasoning, any instance in which "x is y" implies both that the conditional statement: "if x, then y" and its converse: "if y, then x" are both always true together, but the converse of "if born of the spirit, then spirit" (John 3: 6b—see # 1456) is not true using the words attributed to Christ in the Holy Bible. Specifically, since Christ stated, "God the Father is a spirit . . .", according to # 1458 (John 4:24), but cannot be said to have been "born" as # 1459 (John 17: 3) assumes that God (the Father) is eternal in order for eternal life to be directly connected to knowing God (the Father). The general pattern though for such equivalence relationships as employed in the words attributed to Christ in the Holy Bible is that the converse does hold, and the other equivalence relationships that do not result in the converse being true, such as # 1478 in Appendix A, cannot be shown explicitly to be false using strictly the words attributed to Christ in the Holy Bible. {Note: As an intriguing aside, the Gospel according to John contains all but a few of these statements involving equivalence relationships.} Notably, # 1459 in Appendix A is not an exception since "God is a spirit" only implies inclusion or that "If a thing is God, then that thing is a spirit", but not its converse and since 'inclusion' statements do not have the requirement that the converse also be true. Interestingly though, the converse of # 1459 is not true, but then the formal rules of human reasoning would not require it to be true.

4) The transitive property of equivalence relations does not always hold. Specifically, there are three (3) instances in which the transitive rule of equivalences does not hold when this study's assumption of literal interpretations only is applied. This formal rule of human reasoning states that "If A=B and if A=C, then B=C".

First, in # 1446 from Appendix A, the equivalence of "this" is to "my (Christ's) blood of the new testament" in Matthew 26: 28 is not the same via a literal interpretation as given for "the new testament in my blood" from I Corinthians 11: 25 (which corresponds to # 1561 in Appendix A), which is different mainly because of the use of the preposition, "in". This preposition has a variety of meanings in the New Testament books of the Holy Bible (including uses as: in, on, at, within, among, with, by, or by means of), but which literally usage only as per the assumptions of this book results in a different equivalence since the Matthew 26: 28 passage literally implies that the "cup" is "(Christ's) blood . . .", while the passage in I Corinthians 11: 25 literally implies that the "drink" ("cup") is "the new testament . . .". Notably, all of the English translations checked (KJV, NIV, NASB, and Amplified) all translate the prepositional phrase after "new testament" in I Corinthians 11: 25 in the same manner—specifically as "in my blood", but only the King James Version out of this list translates I Corinthians 11: 25 as other than "my blood of the covenant". Since this study assumes use of the King James Version throughout, this 'exception' must noted; however, in comparison to the other three (3) types of 'exceptions' from just above, this 'exception' has the caveat that there is a definite explanation available from the use of the King James Version. Hence, this fourth type of 'exception' is not dismissed, but it is also not emphasized in the analysis that is following for these reasons.

Second, in # 1458 from Appendix A, the equivalence of "know thee, the only true God, and Jesus Christ" to "eternal life" in John 17: 3 is not the same via a literal interpretation as given for "his (God the Father's) commandment" from John 12: 50 (see # 1487 in Appendix A). While this equivalence assumes the same Greek words for "eternal life" and "life everlasting", "know(ing) God (the Father) . . . and Jesus Christ" is not the same literally as "his (God the Father's) commandments". This is a more significant instance than the first occurrence from just above since this difference cannot be reconciled simply by differences in translations of the Holy Bible, and as such this 'Exception' is addressed in further detail in Chapter 4.

GENERAL OBSERVATIONS REGARDING THE NOTED EXCEPTIONS

Of these exceptions, the only part which is not of significant importance technically or theologically is that which can be essentially attributed to the use of English words that are not generally "binary" (that is, the term, "good" in English is accepted to have a range of "goodness" mixed with less than "perfect goodness" and that the opposite of "good" is not "bad"). Within precisely define axioms, such as mathematics or strict human logic of any form, this sort of "binary" nature of terms can be assumed (i.e., a object is either a rectangle or is it is not according to whether or not the definition of a rectangle can be satisfied), but with regard to normal human communication and apparently in Christ's Use of logic it is possible for "(not) good" to be not equivalent to 'bad'.

VIOLATIONS NOTED FROM COMPARING CHRIST'S USE OF LOGIC TO FORMAL HUMAN REASONING

Of critical importance to this analysis is the observation that Christ's use of logic permits a statement to be true assuming only comparison to other statements attributed to Christ in the Holy Bible, but to be false by the strict rules of human reasoning. As such, these are very generally, though not entirely, different than 'exceptions' noted from comparing only the words attributed to Christ in the Holy Bible to each other, and as such, these instances are 'violations' when viewed from the perspective of formal human reasoning only. The structure of the conditional/procedural forms noted as yielding 'viola-

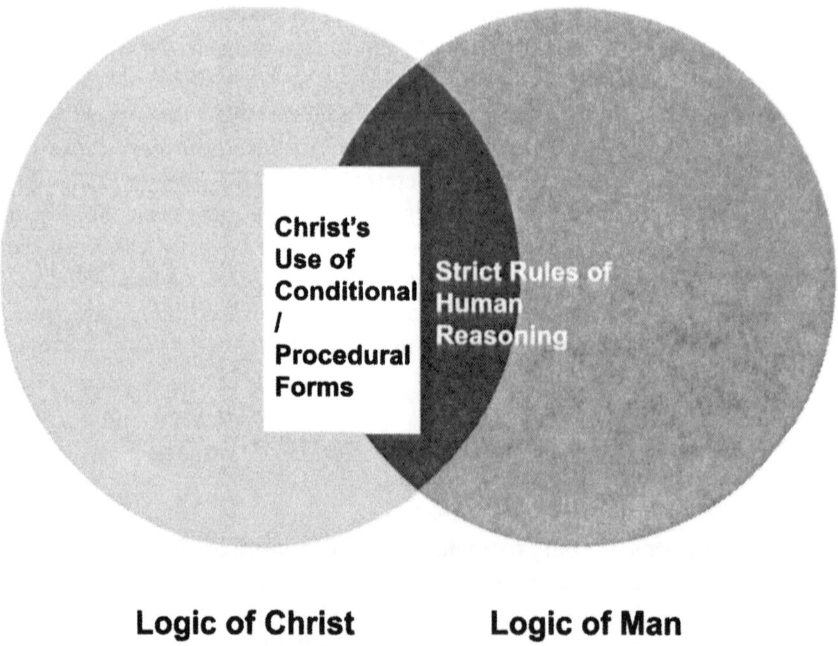

Figure 1.1. Diagram of Relationship Between Christ's Use of Conditional/Procedural Forms and Strict Rules of Human Reasoning

tions' are not essentially different than the conditional/procedural forms which do not have a 'violation', though in a few cases (such as # 536), the condition is a description of God's Character, not a specific action.

There are fifty-one (51) such 'violations'. All but two (2) of these 'violations' are found in the Gospel According to John via examination of conditional statements involving equivalence relationships.

CHRISTIAN AXIOMS

Chapter 5 details the 'irreducible' or 'axiomatic' conditions from which all other implied conditions and implied conclusions/promises in the words attributed to Christ in the Holy Bible result. Fourteen (14) such 'axioms' are identified, listed by frequency of occurrence as a measure of 'priority', and the relationships between these 'axioms' are explored in detail, including specific rationales why particular conditions are not listed as 'axioms' given their being implied by one (1) or more of the fourteen (14) axioms. Finally, the relationships of these 'axioms' to the other primary lists of 'Christian virtues' from the New Testament of the Holy Bible is examined.

In so doing Chapter 5 is seen as an application of the analysis in chapters 1 through 4, though only as one (1) type of application. These foundational or 'axiomatic' conditions for the conditional/procedural statements attributed to Christ in the Holy Bible focus on attitudes, even for conditions involving direct action on the part of a person.

Thus, within these bounds, the identification of the 'unimplied' or 'axiomatic' conditions from the words attributed to Christ in the Holy Bible is a valid exercise logically. Also, it is hoped that such may be practically useful as well in terms of identifying key, spiritual principles for those seeking to think and act according to the teachings accorded to Christ.

Chapter Two

Logical and Analytical Methodology

ASSUMPTIONS

The methods and assumptions employed in this analysis are intended to be technically rigorous and comprehensive within the scope of the words attributed to Christ in the Holy Bible as interpreted only in a literal manner. The rigor of this analysis is such that only seven (7) assumptions are made:

a) All conditional and procedural statements attributed to Christ are treated as 'true' from a logical perspective using only literal interpretations of words attributed to Christ for the purposes of permitting internal and external comparisons, though both 'exceptions' and 'violations' to such treatment wherever found.
b) Only literal interpretations are considered valid for this study since the formal rules of human reasoning do not consider or recognize any other interpretations. (Note: There is one (1) instance in which this assumption is not valid to be applied—specifically in # 909 from Appendix A for "there is none good, but God" since the context of the passage is deemed to warrant a rhetorical, not strictly literal, interpretation, though this is no manner is considered a 'symbolic' or 'figurative' interpretation.)
c) In the analysis for identifying the agents indicated or implied in the conditions or conclusions of these statements, Jesus Christ is assumed to have the same 'agency' as God the Father since Jesus Christ is to have stated "one(ness)" with the Father.
d) Some conditions can only be met by effort involving humans and God and/or Christ, though not all within the first four (4) chapters of this study in light of the forms denoted as 'MAMR' or 'MAGR'.

e) Mankind has a volitional will upon which he/she can choose with literal consciousness and with literal implications to act or not to act even with the power and action of the Trinity present.
f) Human knowledge is epistemologically both declarative (statements of fact) and procedural (if/then as condition/action pairs) and that declarative knowledge is processed by humans into condition/action pairs by matching existing procedural knowledge with declarative information—see John Anderson's Architecture of Cognition (1983). Comprehensiveness in this scope is insured by comparing all of the conditional/procedural statements attributed to Christ in the Holy Bible (both those stated directly and those 'inferred') to each other (internally) and to all of the formal rules of human reasoning (externally).
g) While the number and specific authors of the books in the Holy Bible used for this analysis may be debated, the words attributed to Christ in the Holy Bible are not assumed to have any input in concept or phrasing to these words as otherwise the words would not be solely attributable to Jesus Christ.

LOGICAL METHODOLOGY

The conditional/procedural forms have been determined not just by searches for keywords (such as 'whomsoever', 'if', and 'then'), but rather by a process of studying every statement attributed to Christ in the Holy Bible. Specifically, the conditional/procedural statements have been determined by identifying any statement attributed to Christ that has a specific condition (directly via the use of the words, "if", "since", "because", or "for"; or 'inferred' via descriptions of a general class of persons or things or via definitions pertaining to specific persons/beings) and a conclusion (directly via the use of the word, 'if', or 'inferred' via the same rules as for the associated conditions. The procedural statements have been determined by the same rules as those just indicated for conditional conclusion statements, but limited only to conditions and/or conditions involving a specific time or time period. This is considered especially with statements of the form "When x occurs, then y will happen" (for examples see # 84 and # 317). {Notes: application of this methodology revealed that 82% of the words accorded to Christ in the four (4) Gospels can be considered either "conditional" or "procedural", with none of the books in the New Testament of the Holy Bible that contain such statements having a significantly greater or lesser percentage of statements that are conditional or procedural compared to the overall average of 82%. The word "when" is not used interchangeably with the word "if." Specifically, "when" implies the context of a command or a condition that is predicted by Jesus to occur with certainty, not a condition involving willful choice.}

It should definitely be noted that 99% of the 1653 conditional/procedural forms determined from this methodology must strictly be called "Inferred" since the use of the words ("if", "since", "because", or "for") in Christ's Words in Matthew and Mark is not frequently included in an explicit manner in the translation used (King James Version) for this study. For example, item # 587 reads in the King James Version as follows: "He that believeth not shall be damned". The "inferred" form used for analysis of this item # 587 is as follows: (if one) believeth not, (then that one) shall be damned. Another main example of how "inferred" conditional/procedural forms are derived is found with the use of the English word, "every". Consider item # 65, which in the King James Version says, ". . . every good tree bringeth forth good fruit". The "inferred" form of this item for analysis is as follows: (If a) . . . tree (is) . . . good, then (the tree) bringeth forth good fruit.

Examples of a passages that are not able to be "inferred" as a conditional/procedural form are also evidenced (though much less numerous than those that are directly conditional/procedural or that can be "inferred" as above to be conditional/procedural). For instance, consider the command in Mark 16: 15 in which Christ said, ". . . Go ye into all the world, and preach the gospel to every creature." This statement is not analyzed in this study as a conditional/procedural form since it is considered fundamentally to be a purely declarative statement given that there are no specific conditions or sense of timing associated with the commands to "go" and "preach". Also in this vein, consider, Matthew 28: 18 which in the King James Version states, ". . . All power is given unto me in heaven and in earth." This is also not considered to be a conditional/procedural form for the same reasons.

The only *apriori* assumption (other than that all of the conditional/procedural statements are true from a strictly literal interpretation) are made here regarding the unique aspects posed by Christ's own statements regarding His Humanity and Deity (+) is that Jesus Christ claimed the same nature and attributes as God the Father. No other theological assumptions made. The primary reason for studying the words of Christ in the Holy Bible and not those of others in the Holy Bible is that these words form the largest set of statements attributed as being actually stated, not just written, by a single individual in the New Testament.

(+)—The author of this analysis affirms personal 'belief' in the theological truth of these statements, but importantly the author's or others' personal beliefs regarding the theological/spiritual truth of such statements are not relevant to this study since all analyses here are specific only to internal consistency from only literal interpretations and external comparisons to formal human reasoning, which makes no explicit provision for the concept of 'belief' (i.e., any idea or thing that cannot be proven via repeatable, tangible evidence and/or via the axioms and rules of formal human logic).

ANALYTICAL METHODOLOGY

The analysis of the conditional/procedural statements in the four (4) Gospels attributed to Christ has two (2) primary goals. First, a comparison of the logic that Christ used in His conditional/procedural statements is then made to His other conditional/procedural statements to determine the logical categories and degree of consistency in these conditional/procedural forms. Second, the logic that Christ used in His conditional/procedural statements is compared to formal human reasoning.

In trying to reach both the first and second goal, all formal rules of human reasoning are compared to the conditional/procedural statements attributed to Christ in the Holy Bible. No logical differencing has been used to demonstrate the veracity of any statement, except that of other words attributed to Christ. The only difference between the first and second goals is that assessed 'truth values' for the first goal are based solely on internal consistency within the statements accorded to Christ in the Holy Bible, while the 'truth values' for the second goal are based on literal, explicit matches/non-matches to the formal rules of human reasoning only.

So, for example with the first goal, if Christ is to have said "if P, then Q", then the only manner in which its contrapositive, for example as a logical equivalence, is stated as 'true' is if one of the other statements attributed to Christ in the Holy Bible states or is inferred to state (via the rules for 'inferred' statements in the logical methodology section) that if NOT Q, then NOT P (with the exact same wording, not just similar). To show a contrapositive to be false will require a direct statement in the words attributed to Christ in the Holy Bible to that same effect or a direct counterexample with the same wording exactly for the condition (NOT Q), but a resulting conclusion of NOT P (with same wording exactly) being false. More precisely, the contrapositive will be seen as 'false' when (NOT Q) is true but (p) is false (which is the only condition under which any conditional/procedural statement can be judged to be 'false', as all other combinations necessarily produce an overall true statement even though the cases in which the condition of the statement is false are said to be "vacuously" true overall). Any 'false' truth values from such comparisons are referred to as 'exceptions' (as opposed to 'violations') since, while the same rules of formal human reasoning are employed, no *apriori* assumption regarding the structure or relationships between the logical statements are made for the words attributed to Christ in the Holy Bible, nor indeed necessarily that Christ employed a definite, axiomatically based set of rules for logical differencing.

The second main goal of this analysis is simple to state, though it has been the most time consuming part of this overall analysis. Specifically, the conditional/procedural forms attributed to Christ in the four (4) Gospels are compared directly with all of the formal rules of human logic. In so doing, the basis for comparison with this second analytical goal is the formal set of rules upon which all scientific, logical efforts are based since no overarching set of rules is assumed *apriori*. Truth values are assessed based on 'violations' found from statements or multiple statements taken together which are not consistent with any of the formal rules of human reasoning.

TECHNICAL NOTES REGARDING THE FORMAL RULES OF HUMAN REASONING

The formal rules from human reasoning are taken from a mathematics textbook (Rosen, Kenneth H., "Discrete Mathematics and Its Applications", Fourth Edition, WCB/McGraw-Hill, 1999).

Human use of what is often called "common sense" is strictly different from the formal rules of human logic and as such "common sense is not addressed here. Some examples from human "common sense" that show such aspects are:

a) Opposites attract; birds of a feather flock together.
b) Too many cooks spoil the stew; many hands make light work.
c) Absence makes the heart grow fonder; out of sight out of mind.
d) The greasy wheel gets the grease; attract more flies with honey than with vinegar.

DESCRIPTION OF NOTATIONS/ABBREVIATIONS

The contrapositive of a statement is written with a question mark (?), if unsure as to its truth. (Biblical references are shown if given in Bible directly).

Analogy- Analogy (literal)—See Note just below
Contrast- Contrast

Statement, Inverse pairing in near proximity in context- STINV
(Inferred if, then form)= (INFERRED)
Man's actions/attitude, God's result in future- (MAGR-F) - though for man in many cases
Man's actions/attitude, God's result now- (MAGR-N) - though for man in many cases
God's Action, God's Result- (GAGR) - though for man in many cases
God's Action, Man's action/attitude- (GAMA)
Statement addresses a specific question raised by listener- concrete (P)
Quoting Old Testament- (OT-Q)
Referring to Old Testament- (OT-R)
Contrapositive seems true, but direct verse needed- verse?
Contrapositive seems false, but direct verse needed- verseF?
Speaking to multitudes (Mult)
Speaking to disciples (Disc)
Speaking to apostles or a subset of apostles (Apos)
Speaking to educated individual (Indiv-E)
Speaking to uneducated individual (Indiv-U)
Speaking to religious leaders (Relig)

PRIMARY LIMITATIONS OF ANALYSIS/NOTATIONS

1) No allowances for the possibility of multiple 'truth values' is permitted since only literal meanings are employed in assessing 'Christ's Reasonableness' here and since such is not consistent with the formal rules of human reasoning, though the implications of such in terms of attempting to find common patterns and underlying factors for the noted 'Exceptions' is documented in chapter 5.

2) Matching for conditional/procedural references is from an English translation, not from the original Aramaic or Greek. The degree to which subjunctive forms and conditional forms are used in other translations or even in Aramaic or Greek is not considered here. Notably though, randomly selected passages from The King James Version when compared to three (3) other major translations (NIV, NASB, and RSV) did not reveal substantively different uses of conditional/procedural forms.

3) The overall total number of conditional/procedural forms noted (1629) in Appendix A does not account for twenty-four (24) additional procedural forms in which the condition involves "when", two (2) additional conditional forms in which the condition involves "since", three (3) additional conditional forms that have an "if . . . then . . ." form given that these twenty-four

(24) additional forms are enumerated by numbers and letters, such as 77a) for instance. This revised total number of conditional/procedural forms would be 1653. This total does not account for replications of the same or very essentially the same statements, for which the total number of unique conditional or procedural statements is 1396.

Chapter Three

Comparative Analysis

The following observations and categorizations are based on specific analyses of all of the words attributed to Christ in the Holy Bible both internally with each other and externally with all of the formal rules of human reasoning.

The passages examined indicate that these conditional/procedural (Conditional/Procedural) forms (i.e., "if p, then q" forms either directly or inferred) have significant overlap with strict rules of human reasoning. Specifically, thirty-two (32) different logical forms have been identified, most of which correspond directly to the logical forms used by the strict rules of human reasoning. A total of 1396 unique conditional/procedural statements have been noted.

TYPES OF CONDITIONAL AND PROCEDURAL FORMS

As regarding the most common forms noted in this analysis, the most consistently noted type of conditional/procedural form is from the 101 definitions/equivalence relationships noted in Appendix A from # 1602-1611. This set of 101 definitions does not include the seventeen (17) dual-inclusions, which would imply according to formal human reasoning an equivalence relationship, but these dual-inclusions are not added to the list of definitions/equivalence relationships since all of these dual-inclusions involve 'violations' of formal human reasoning based on their implication via literal interpretation that Jesus Christ and God the Father are 'equal' and that Jesus Christ and the set of all disciples of Christ, Jesus Christ and the eleven (11) Apostles, or even Jesus Christ and an individual disciple are also 'equal'.

(See details on this form in the section below in Chapter 3 regarding "converses" and in Chapter 4 regarding categorizations of the noted 'violations' and 'exceptions'.)

The 2nd most consistently noted type of conditional/procedural forms are the seventy-six (76) logical forms of the following type: A implies B (with) since C implies B. This "explanatory" type of condition is important in the sense that "since" or "because" is not used significantly elsewhere and given that this logical form provides additional insight, not an additional condition. This form would not necessarily be in exception of strict rules of human reasoning even if an additional condition were used here, but since the statements following "since" are strictly of an explanatory nature, this situation does not have to be considered. See the section below in Chapter 3 regarding 'Comparisons Between the Books Analyzed' for more details.

The 3rd most consistently noted conditional/procedural form is of the type: "if p, then q" and "if NOT p, then NOT q" (which is called "statement and inverse"). There are fifty-one (51) of these "inverse" forms, which are noted as statement/inverse" (STINV) pairs. While much more detail in addressed later in Chapter 3 to this "STINV" form, it should be noted that the strict rules of human reasoning do not require that a statement and its inverse be simultaneously true or false together. In Christ's use of logical forms, no instances are noted in which a statement is true with the inverse being false, though in extending beyond the conditional/procedural forms noted in this study, words attributed to Christ do permit one (1) instance in which the inverse of a statement can be false while the statement is true. (See Matthew 5: 7 and Matthew 5: 45)

Very notably, the remaining forms noted here encompass the complete set of all formal rules of human reasoning, except for the logical forms of 'transitivity for equivalence relationships' and the conjunctive forms involving 'or' (i.e., A or B) as a condition or as a conclusion. None of these other forms have large numbers of occurrences relative to the three (3) main forms noted above, but the observation that all but these two (2) types of forms are noted in Christ Use of logic is highly significant.

DEDUCTIVE AND INDUCTIVE FORMS

In all of this analysis, the concepts such as "often" and "sometimes" are seldom used in Christ's Use of logic, but the concepts of "always" and "never" are very common. The implication of such qualifiers is that Christ's Use of logic is substantively similar to strict rules of human reasoning in this specific regard. There is no definitive pattern here though since Christ used terms like "could" and "might" not infrequently, which suggest more the concept of non-universality of a particular condition.

Moreover, the nature of the logical forms used by Christ is deductive very generally, though there are several instances that are specific to a particular context or particular person. By deductive reasoning, it is meant that the statements attributed to Christ very largely follow a pattern that is not specific to individuals, nor specific times or time periods, nor is the reasoning from specific instances or sub-groups to more general conditions, and nor is the reasoning of a progressive, continuing nature. In formal human reasoning, a good example of deductive reasoning would be: "if an object is a triangle with all sides equal, then all of its interior angles are equal as well." In the words attributed to Christ in the Holy Bible, a solid example of deductive reasoning can be found in Luke 16: 15 (see # 1453 in Appendix A): "(If a thing) . . . is highly esteemed among men . . . (,then that thing) . . . is abomination in the sight of God."

Three (3) types of inductive reasoning are employed in these conditional/procedural forms. There are 153 of these inductive forms in the statements attributed to Christ. Specifically, the first type of inductive reasoning has two (2) examples. These two (2) special examples are found in # 1362 (see John 17: 7-8) and in # 1375 (see John 17: 20-21), and are inductive in the sense that a specialized, "subset of the general conclusion" is part of the condition with an additional condition, such as "receive Christ's (Words)" or the disciples being "one is us (God the Father and Jesus Christ), being required for the more general conclusion to be true. The second type of inductive reasoning has a condition that is specific to a smaller group than the associated, generalized conclusion. The one (1) example of this second type of inductive reasoning is found in # 1093 (see Luke 6: 45a). The third and much more frequent type of inductive reasoning is demonstrated by examples in Luke 9: 48b (see # 699) and in Matthew 7: 1-2 (see # 1532). Statements of this first type are only conditional and not procedural statements since the times involved are not specific to a known time or time period. This third type of inductive reasoning, of which there are a total of 150 specific instances, is characterized by progressive, continuing action on the part of humans in both the condition and conclusion, where most have conclusions that "shall/shalt' or "will" (be or occur) once a condition is realized, though at an indefinite time.

Even more specifically, while the vast majority (81%) of Christ's Use of conditional/procedural forms have general conditions/conclusions applying to all people, 181 examples have been found in which a conditional statement appears to be specific to a particular individual (for example, the "rich young ruler"—see item # 206; as well as the woman with the issue of blood and

Jairus—see 2nd note after # 688). The form of reasoning by which these cases are labeled is "Neither inductive nor deductive". In all but one of these cases where a specific individual(s) is(are) named in the condition of a conditional/procedural form, the conclusion is specific to that particular person(s), including for the members of the Trinity and for Satan. The one (1) exception to this general rule is the specific case of # 698 in which the conclusion is general, yet the condition appears to be specific to just the disciples within the context of the verse—see Luke 9:48c.

It should also be noted that there are a total of 246 conditional/procedural statements (5.6 % of the unique statements) attributed to Christ that are neither inductive nor deductive since they do not fit either the 'inductive' or 'deductive' definitions strictly. The Gospel According to John contains a very significant majority (150) of these instances, which is very understandable given that roughly 25% of the conditional/procedural statements in the Gospel According to John are specific to Jesus Christ. More generally, all of the statements of this type are either specific to a particular individual or a particular time/time period, though there are a few examples of statements made to an individual person which are considered to be "inductive", such as # 1265 in Appendix A, which while stated to and involving Saul/Paul the Apostle, is "inductive" since it is "through the witness" of Saul that the conclusion "may" be realized and as such indicates the beginning of a process of "opening the eyes (of the Gentiles)".

THE NATURE OF THE CONDITIONS AND CONCLUSIONS IN THESE CONDITIONAL/PROCEDURAL FORMS

Examination of the nature of the conditions and conclusions within the conditional/procedural forms attributed to Christ is also needed. The analysis begins with examination of the conditions associated with these forms.

Specifically, there are 1396 unique conditional/procedural forms (which is approximately 84 % of the total of 1653) noted in Matthew, Mark, and Luke, and John. Approximately 55% (759 out of 1396) of the total number of unique conditional/procedural forms have identifiable conditions exclusively denoted as either 'MAMR' or 'MAGR' involving "man" primarily (with 282 instances having a conclusion focused on God's Action/Result, which are referred to by the notation of "MAGR"). Notably, the "GAGR" forms (where both the condition and conclusion specific to God the Father, Jesus Christ, and/or the Holy Spirit) are the lowest overall in comparison to the other three (3) forms (with a total of 201 forms of this type noted overall). As a result, the "MAMR" forms are the most frequently occurring type of logical form overall, with a total of 477. Also, some conditions and conclusions involve both God and man, so there are overlapping cases as well. It must be noted that such conditions are described as 'exclusively denoted' since there are conditional and procedural statements in which Jesus Christ and/or God the Father are indicated or implied besides solely a human or group of humans even within this set of statements for which humanity is part of the condition (see forms of the type 'GA-MA/MR', 'GA-MA/GR', or 'GA-MA/MR-GR'. As indicated by the preceding notation, an essentially similar effect with both is noted for the conclusions in regard to both 'man' and 'God' being indicated or implied. (Note: There are only fourteen (14) instances in which no designation regarding the type of condition or conclusion is made.)

With regard to the various audiences to which Christ is indicated in the Holy Bible as having spoken these conditional/procedural forms, there does not appear to be a significant difference depending on the audience, though it is stated in Matthew 13:34 and Matthew 13: 10-16 that Christ only used analogies/parables with the general audiences (non-disciples) so that "seeing they might not see . . ." and that toward the end of His Presence on earth that He stopped using parables with the disciples. Also, it should be noted that even in speaking to the "multitudes" that Christ used all of the four (4) primary conditional/procedural forms (MAMR, MAGR, GAMR, GAGR) and approximately similar proportions for each of the types noted in the words attributed to Christ when His speaking to the "multitudes" is compared to the overall percentage of each form in all 1653 items analyzed. In speaking to the religious leaders and with the disciples, essentially similar patterns occur, though for the conditional/procedural forms in which Christ's Audience included the disciples the use of the "GAGR" form is somewhat more common than would be expected in comparison to overall usage of the "GAGR" form, especially since it is the least frequently employed type of logical form.

It must be noted that these trends do not account for the forms related to analogical contexts (including those forms taken from the parables since parables appear to permit meanings beyond just a literal interpretation). However, in the parables, if the "literal" meaning is only considered (which has been stated as an assumption in this study), then the form "MAMR" is significantly used by Christ. So, in the parables, Christ often used forms such as item # 212: "when he (an householder) had agreed with the labourers for a penny a day, (then) he (the householder) sent them into his vineyard".

Another interesting observation is that the conditions or conclusions here do not significantly employ the use of the Greek word(s) translated into the English word, "or" (via the King James translation). While conjunctive-type uses of "or" appear fairly frequently in the conditions of the conditional/procedural statements of Christ and only five (5) times in the conclusions of the conditional/procedural statements of Christ (see # 304-306, # 889, and # 891), there are only two instances in which Christ's Words employ a "choice of action" via the use of the word, "or" (see # 682 and # 782). This would seem to suggest that Christ's Use of Logic is very definitive with little room for optional conclusions (that is, if A, then B OR C) appears only twice). Comparatively, the use of the Greek word(s) translated into the English word, "and" (via the King James translation) occurs very frequently in both conditions and conclusions so that Christ's Use of Logic permits significant use of the "layered" conditions and conclusions in which multiple conditions imply a conclusion and/or a single or multiple conditions imply multiple conclusions.

CATEGORIZATION OF CONDITIONAL/PROCEDURAL GENERAL FORMS

Having described the nature of the conditions, conclusions, audiences, and deductive patterns within the conditional/procedural forms attributed to Christ, we now provide a categorization of all of these forms (with those in the bolded text are instances where definite 'exceptions' as defined previously are noted), where the basis for these categorizations is found from generalizing all of the 1653 forms noted in Appendix A. There are a total of thirty-two (32) logical forms noted from this analysis.

GENERAL NOTES

1) NOT (NOT A) does not equal to A in all cases since NOT A sometimes permits "neutral" condition to also still be true (such as seen especially in STINV cases like NOT "break one commandment" in # 15 is not the same as "do and teach" since one can both "NOT break" and "DO" at the same time given "omission" traits), though in other cases Christ even uses the same condition to imply opposites (such as "have everlasting life" and "not perish/not condemned" with the condition of "believeth on him" in # 1023).
2) NOT (A AND B) is treated as NOT A OR NOT B.
3) IF A, THEN B with If B, THEN C (some of these are sequencing of events in time)
4) IF A, THEN B with If A, THEN C (also (A AND D implies B with A AND D implies C)—see # 1023 and # 1056 ; see also IF A, THEN B, with IF A AND D, THEN C, where B has the same concept in meaning though in

different words than C (though B does not imply C nor does C imply B in Christ's Words strictly), such as with # 1089 and # 1090 and also # 1023 and # 1090

5) IF A, THEN C with IF B, THEN C (# 96 and # 479 and #354 and #361, # 16 and # 511); see # 788 and # 789 for instance of where B is a special case of A so that A implies B); also A−>C, B implies NOT (NOT C) #171 and #475
6) IF A, THEN B with IF NOT A, THEN NOT (see STINV statements in Appendix A; "statement and inverse", which are the most frequent type of conditional statement pairs noted in this analysis, and as such are juxtaposed as "chiasms")
7) IF A, THEN B, IF NOT B, THEN NOT A ("statement and contrapositive"—only 1 'exception' of this type found using the words attributed to Christ, though many such noted more generally—(see #1045 and #1396)
8) IF A, THEN B with IF NOT A, THEN C (see #84 and 96)—see also A implies B and C, NOT A implies D and C. (# 709 and # 710) and (# 84 and # 712)
9) IF A, THEN B with IF B, THEN A ("statement and converse"—Mark 7:9 and Matthew 15:3; John 14: 21, 24; see # 1032 and # 650 for close to converse). See # 1081, # 1087, and # 1089 for examples of definitions, though the converse is not noted explicitly).
10) IF A, THEN B with IF NOT B, THEN C (#470 and #93 close and # 1001 with # 233; also # 1021 and # 475)
11) If you do p, then q shall happen to you (Personalized to an individual—see Matthew 19:21 and Mark 10: 21 for rich young ruler as these words do not seem to apply to all individuals)
12) If A AND B, THEN C.
13) IF A, THEN B AND C.
14) IF A AND B, THEN C as well as IF A AND D, THEN C (See # 279 and # 761 and see # 497 and # 201)
15) IF A AND B, THEN C as well as IF A AND D, THEN C (See # 233 and # 1254) with If A, Then E (# 52)
16) If A, then B and C as well as IF A, THEN C (See # 187 and #846) ; see also IF A−> D and A−> B with A and E −> B and D (see numbers # 1023, 1029, and 1056)
17) If A, then B and If C, then NOT B (See # 811 and # 1023)
18) If A, THEN B and IF B, THEN NOT A. (See # 1118 and #1037)
19) If A, THEN B and IF C, THEN A (See # 1118 and # 1083
20) If A, Then NOT A along with IF A, then B (See # 97, # 480, and # 1210)
21) If A and B, THEN C, NOT D, and E with IF A and NOT B, Then D (see # 1056 with # 1215)

22) IF A, THEN B; IF A and D, THEN B; IF A and C, THEN very similar to B; IF E and F, THEN A and B (SEE #57, #233, #1254, AND # 1290); (see also # 1254 and # 1300); see also # 233 and # 1348 for (IF A, THEN B; IF A and D, THEN B)
23) IF C, THEN D with IF C, THEN A and IF D, THEN B, and more generally IF A, THEN B (see # 1059 and # 1333)
24) IF A, B, and C, THEN D with IF A, then NOT E where E and have the same concept of meaning (see example in # 521 and # 890)
25) A 'is/am/be' B (see # 1458 for a third type of 'Exception' involving dual inclusion from equivalence relationships
26) A is a (member of) B (see # 1459 and, for the negative form of membership, see # 1547)
27) If A, then (a person may, not necessarily will do) B—see # 1182 & # 1183
28) If A=B and if A=C, then B is not necessarily equal to C—see # 1446 and # 1561; and # 1458
29) (since) A Then B, and (if) B, then C, and If C, then D—see # 1323 and # 1324 for an example
30) (if) A, Then B, and (if) B, Then C, and (if) A, Then C—see # 1242, # 1023, and # 1056 for the one (1) example of this form
31) (since) X is a subset of Y and (since) Z, then Y—This is a key example of inductive reasoning in these forms.
32) (if) A, then B with C is or does A, implies (if) C, then B—see # 1086 and # 1087 and # 1283 and # 1284) for examples of syllogisms

COMPARISONS OF THE CONDITIONAL/PROCEDURAL FORMS BETWEEN ALL BOOKS ANALYZED

The conditional/procedural statements attributed to Christ are highly similar by all measures employed in this analysis. In fact, the only key difference is a result of the significant number of definitional statements (equivalence relationships) attributed to Christ in the Holy Bible and in this case primarily about Christ's Nature.

This high degree of similarity includes the total number of conditional/procedural forms, except that The Gospel According to Mark is significantly lower in terms of total number of conditional/procedural forms. This difference though is not deeply remarkable since the total number in Mark's Gospel is very close to being proportional to the other Gospels given that its relative number of forms attributed to Christ in Mark's Gospel is close to half as large as the proportion of conditional/procedural forms to the total statements overall in the Holy Bible.

The Gospel According to John has the highest total number of conditional/procedural forms at 459. The Gospel According to Luke has the 2nd most conditional/procedural forms of the four (4) gospels with a total of 428 forms, though only slightly in comparison to The Gospel According to Matthew, which has a total of 409 conditional/procedural forms. The Gospel According to Mark has 124 of its total of 196 forms that are either the same or very similar to one in The Gospel According to Matthew. The Gospel According to Luke has 145 of its 428 with overlap in either or both of Matthew or Mark's Gospel, and The Gospel According to John has only 66 of its 459 conditional/procedural forms with overlap in Matthew, Mark, or Luke.

There are eight (8) significant differences noted from this analysis between the four (4) different Gospels involving the conditional/procedural statements attributed to Christ. These differences are focused clearly on the Gospel According to John and result strongly from the author's stated purpose of demonstrating the Deity of Jesus Christ.

The first difference involves the extensive use of definitions in the Gospel According to John (see statements 1399–1476 in Appendix A). These definitions also all employ the use of various forms of the verb which translates into English as, "to be" (specifically 'am', 'is', 'are', and 'be'). The associated issues that result from such significant uses of the definitional forms occur when literally connecting Christ as a person to inanimate objects ('bread', 'water', 'drink') or to God the Father via Trinitarian statements. See the notes in Appendix A for extended comments on specific instances. The definitions are referred to as equivalence relationships in the formal rules of human reasoning.

The second main difference is that the vast bulk {all but two (2)} of the 'violations' and most of the 'exceptions' noted from this analysis occur are connected to these equivalence relationships or definitions from the Gospel According to John. Thus, the first main difference along with the nature of the definitions being specific to various aspects of the Trinity is the 'real' difference when comparing the four (4) Gospels, with the second difference being an 'effect' of the first difference.

The third main difference between these conditional/procedural forms is that The Gospel According to John has just over 60% of the total number of conditional/procedural forms of the following type: If A, then B (with) since C implies B. The instances of this type involving this form with "since" (where "since" also is employed as a general term for "since", "because", or "for") in The Gospel According to John are not though fundamentally different in nature from the instances where this form occur elsewhere. This is so because this particular form is used exclusively to provide additional detail for the condition, but not a separate condition. There is one (1) interesting pattern that develops in the uses of this form from John's Gospel. Specifically, most of the uses of this form in John's Gospel where 'C' (referencing the general form from just above) is an explanatory statement specific to the nature of Christ, especially His relationship and actions relative to God the Father. This is not considered remarkable though since the overall theme of John's Gospel is focused on the nature of Christ's Deity.

Fourth, approximately 60% of the total number of conditional/procedural statements that employ a form of reasoning that is considered 'neither inductive nor deductive' are found in the Gospel According to John. This is though quite understandable given the purpose of John in demonstrating the Deity of Jesus Christ and in so doing to address statements about the specific, personal attributes and character of Jesus Christ. Notably, from the books of the Holy Bible analyzed that are not part of the Gospels, a majority are considered to be 'neither inductive nor deductive'. This is a direct result of the fact that a majority of the conditional/procedural statements are particular to two (2) particular individuals—specifically, either Saul/Paul the Apostle and/or Jesus Christ.

Fifth, all of the statements attributed to Christ that involve an inclusion (one-way or dual) are in the Gospel According to John. See # 1495, # 1496, and # 1612-#1626. These statements are considered conditional statements since logical forms where "A is in B results formally in 'IF A, then B". Such statements are useful for demonstrating an assumed sense of non-physical equality of Christ to God the Father and of Christ to the disciples in the sense of the logic employed in the words attributed to Christ. In so doing, however, such statements further highlight the Trinitarian issues mainly found in John's Gospel.

The sixth difference is also very closely associated with the fifth difference in that the statements of inclusion are predominately specific to Jesus Christ or God the Father. This sixth difference is associated with the observation that John's Gospel is much more focused on statements particular to individuals, primarily to Jesus Christ given the stated purpose of John's Gospel in demonstrating that Jesus is the Son of God. Specifically, 117 (65%) out of the total of 179 statements specific to a particular individual in the words attributed to Christ are found in the Gospel According to John.

The seventh difference results from noting that all but one (1) of the 'non-sequitirs' according to formal human reasoning occur in the Gospel According to John. Such occurrences are not part of the formal analysis of this study since such are not considered to be valid from formal human reasoning, but it is noteworthy that the Gospel According to John contains the significant preponderance of such occurrences, of which importantly none are directly tied to the chief issue pertaining to the Deity of Jesus Christ in John's Gospel.

Finally, only one (1) other significant new aspect is identified from inclusion of other statements attributed to Christ from books other than the four (4) Gospels. Specifically, the fourth type of 'exception' noted for transitivity of equivalences is noted (see # 1446 and # 1561 in Appendix A), which is documented for the sake of thoroughness given the use of the King James Version of the Holy Bible for this study.

DETAILED ANALYSES OF MOST COMMON CATEGORIES OF CONDITIONAL/PROCEDURAL FORMS

A. Negating the Negative is NOT always a Positive: In Christ's Use of Conditional/Procedural Forms, it is not necessarily the case that NOT (NOT A) = A as is always true as a strict rule of human reasoning). For example, consider procedural forms # 146 and # 147 which cannot be viewed as the form "statement and its inverse" strictly speaking because "goeth into one's mouth" is not the strict negation of "cometh out of one's mouth" as an object at the same time in reality may be neither "coming out of one's mouth" nor "going into one's mouth" . This lack of strict negation is not always seen in Christ's Use of Conditional/Procedural Forms, for consider procedural forms # 45 and # 46 in which the form of a "statement and its inverse" appear to hold strictly given both the condition and promise of forgiveness from God.

The "middle ground" that the instances in which strict negations do not appear to hold in Christ's Use of Conditional/Procedural Forms would seem to open the door to a non-binary world to consider. This "middle ground" though has very significant implications in that, without being able to reverse negations back to the original statement results in many of the strict rules of human logic not to hold strictly speaking. For example, DeMorgan's Laws of negating "and" as well as "or" conjunctive statements (such as in item # 16's condition of "do and teach") do not hold strictly speaking. Furthermore, this implication goes to the heart of all of the formal rules of human reasoning in that all of the formal rules of human reasoning are dependent upon double-negations holding and its correlate of not permitting multiple 'truth values'.

B. Statements and Inverses: The "statement and its inverse" form (sometimes referred to as chiasms) attributed to Christ requires detailed explanation. In the strict rules of human reasoning, a statement and its inverse are not logically equivalent. That is to say, a statement and its inverse need not be true or false at the same time. For example, in human reasoning, "if an object is a square, then it is a rectangle" is a true statement in general, but the inverse is not (namely, if an object is not a square, then it is not a rectangle" is not a true statement in general (for consider that a rectangle does not have to have all of its sides equal as does a square). However, especially definitional statements are such that the statement and its inverse are true simultaneously. For example, in human reasoning, "if one is a boy, then that one is not a girl" is true, and its inverse, "if one is not a boy, then one is a girl" is also true.

Christ used this type of conditional/procedural form with the "statement and its inverse" as the second most common of the logical forms attributed to Him in the Holy Bible, but He did not always use statements and their inverses as logical equivalents (that is if a statement is true, then automatically the inverse must be true or false necessarily). In terms of general patterns, it might be tempting to make this conclusion. Indeed, no instances of conditional/procedural forms have been found in which Christ said "if p, then q" is true, but "if NOT p, then NOT q" is false. However, two passages in the Holy Bible (one of which is not a conditional/procedural form) that are very similar to a "statement and its inverse" show the statement to be true, but the inverse to be false. Namely, consider Luke 11:10 ("if one asketh, (then) one receiveth") (the inverse is false by Matthew 5:45 as even the unjust still receive), but note that the contrapositive seems true—see #521. To argue then though that Christ intended any procedural (if, then) statement He made to be assumed to be true as well as automatically for its inverse form to be true in all cases would be to argue from silence.

There is another aspect of the "statement and inverse" form to consider. In human logic if the statement and its inverse are both given as true which they are in many places in Christ's Use of logical forms, then this means that the condition and the conclusion are essentially equivalent. For example, in human logic, "if a person is a male, then that person is not a female" is true, but its inverse is also true (namely, "if a person is not a male, then that person is a female). Thus, not being a female is equivalent to being male for humans. This cannot, however, be assumed in Christ's Use of conditional/procedural forms (unless we argue from silence).

The chiastic forms ("statement and its inverse") overall appear to be used to state with very strong general pattern that if the condition of the original statement is not met then the conclusion of the original statement is not met either. Interestingly, the chiastic form is, in all but two cases noted in Matthew and Mark, of the form where that the condition in both the statement and its inverse appears to be mainly focused on man's action or attitude in order for the conclusion to hold.

Thus, mainly with the chiastic form, Christ appears to be saying if mankind does or is a certain thing, then other men or God will do a specific action or have a specific attitude. This strong sense of a condition implying a conclusion and the negative of the condition implying the negative of the conclusion (or conditional reasoning) is replete in Christ's Use of conditional/procedural Forms (second only to "linked logic" in number of instances). Before moving on from this common logical form in Christ's Use of conditional/procedural Forms, it should be noted that Christ did not always add further conditions to all statements He made in such a way that the inverse or a "close inverse" can be assumed in all cases. In two (2) such "statement and inverse" pairs, there are other conditional statements in the Holy Bible which have the same conclusion as at least one of the conditional statements in the "statement and inverse" pair, yet which also has a different condition (see # 45, 46 and # 416 for strong examples).

This demonstrates that the chiastic form cannot be viewed globally to exclude all other conditions resulting in the original conclusion of a "statement and inverse" pair. Thus, when combined with the general form of A implies B and B implies C, this fact suggests that conclusions can be reached with multiple conditions, though the chiastic form still remains the second most common logical form in the Holy Bible.

Notably, Christ, in logical form (# 1055) says, ". . . he who honoureth not me honoureth not the one who sent me". This statement is a key specific example of an inverse since it is very similar in concept to the inverse of the statement in question for 'Exception 1.d)'. This 'inverse-like' statement is not problematic in terms of the strict rules of human reasoning, but is interesting to note since according to this statement (# 1055) the dishonoring of Christ does not imply actual dishonoring of Christ (which is another form of the common 'double-negative' issue), but rather implies dishonoring of God the Father, which is very notable in its own right since this demonstrates explicitly that Christ employed explicitly the use of the contrapositive form (see below in section D).

C. Statements and Converses: One (1) instance in Matthew and Mark indicate that a statement and its very close converse are both given as true (see # 142 and # 455) in Christ's Use of Conditional/Procedural Forms. In human logic, to reuse our continuing example, a statement, S, such as "if an object is a square, then the object is a rectangle" has a Conditional/Procedural converse, in which the conclusion becomes the condition and vice-versa so that the converse of S would be "if an object is a rectangle, then the object is a square". We can readily see from this example that the converse of a statement of a statement and its converse need not be true or false together according to strict rules of human reasoning. However, it should be noted that there are example that occur in human reasoning in which a statement and its converse can both be true. For example, consider another statement, M, as follows: "if an object is a rectangle with all sides equal, then the object is a square". Then, the converse of M is true since a definition of a square is that it is a rectangle with all sides equal. For Christ to have used the converse to be true and the original statement as true is significant as it demonstrates the use of another major form of human reasoning. It should be noted that many people reason (inaccurately) that the converse of a statement must be true if the original statement is true, but if strict rules of human reasoning are applied, it has already been demonstrated above that this need be the case. This would then seem to give further evidence to the idea that there is overlap in Christ's Use of Conditional/Procedural Forms and strict rules of human reasoning.

There is though definite evidence that, using the words attributed to Christ in the Holy Bible, not all converses of original statements must be true. Specifically, there is an instance found from examining equivalence relationships, where "x is/am/be y". Such equivalences would require that both the conditional statement of "if x, then y" and its converse, "if y, then x", be true together. From John 3: 6b ". . . that which is born of the spirit is spirit." (see # 1459 in Appendix A), the words attributed to Christ in the Holy Bible demonstrate that a thing (specifically God the Father) may be a spirit, but yet not be "born of the Spirit". From the perspective of formal human reasoning, this type of 'Exception' is very much like an application of Exception 1) since in order for such to occur, there would have to be an object that is both 'X' and 'NOT X' at the same time given that equivalence relationships according to formal human reasoning imply a bi-directional (both an original conditional statement and its converse) statement based on what is referred to as "dual inclusion" (that is, all members of the set X are in the set Y and vice-versa). The lack of equivalence relationships holding in all cases though is seen as a separate type of exception since no formal rules of human logic involving negations are strictly speaking required to assess the 'truth value' of an equivalence relationship (nor of an original statement and its converse).

Lastly, in regard to these equivalence relationships, it should be noted that there are no explicit instances of the words attributed to Christ in the Holy Bible demonstrating the 'transitive' property of equivalences (that is, where A=B and B=C results in A=C), though there is the one (1) instance in the fourth type of Exception were this rule is not true in the words attributed to Christ. This is significant since there are a number of definitions specific to essentially similar topics, including for "my Father's will" and "the bread of life". There are, however, one (1) noted use of the 'transitive' property of conditional statements (that is, where conditional statements are 'strung together' so that if A then B; if B, then C; implies that if A then C) in conditional statements attributed to Christ (see John 14: 1, John 3: 16, and John 5: 23 in # 1242, # 1023, and # 1056 from Appendix A). A form similar to the transitive property of conditional statements is seen as follows: if A then B; if B, then C, via # 1156 from Appendix A, but this similar form is not exactly the same as the formal rule for the transitive property of conditional statements.

D. Statements and Contrapositives: One other main procedural form needs to be explained to date. In human logic, a statement (if p, then q") is ALWAYS true or false together with its contrapositive. For example, "if an object is a square, then that object is a rectangle" is Conditional/Procedurally equivalent to its contrapositive, "if an object is NOT a rectangle, then that object is NOT a square". While there have been many instances where strict human logic would indicate that the contrapositive of a statement is false (see notation given as "verseF?" to indicate only that the contrapositive appears to be false by strict rules of human reasoning, and likewise "verse?" to indicate that the contrapositive appears to be true by strict rules of human reasoning), it must be noted that no instances of a contrapositive being false while the original statement is true have been noted in Christ's Use of Conditional/ Procedural Forms.

However, it should also be noted that when strict rules of human reasoning are applied, the contrapositive (which, in human reasoning, is always a logical equivalent to the original statement) appears to be false in some cases. Specifically, see John 14: 21 and John 14: 24. This is particularly seen in instances, such as item # 514, in which the negation of the conclusion being true necessitates (by strict rules of human reasoning) that the negation of the original condition be false. In such cases, according to strict rules of human reasoning, the contrapositive would be false since the only time that a "if, then" form is considered to be false is when the "if part" is true, while the "then part" is false. So, while we have seen no instances of Christ directly indicating the contrapositive of a statement being false and in fact only two (2) specific uses of the contrapositive by Christ, strict rules of human reasoning suggest that some of Christ's Use of conditional/procedural Forms do not hold with logical equivalents, such as the contrapositive and negation.

There are several forms in which if taken literally and by strict rules of human reasoning, the contrapositive of specific conditional/procedural statements would *seem* to be false. Most generally, this seems to apply to the forms called "MAMR" in which a man's action or attitude conditions another man's action. Consider items # 885 and # 874 as key examples of this type of instance in which the contrapositive would seem by strict human reasoning to be false while the original statement is true, but not explicitly indicated as such by a literal rendering of the words attributed to Christ as in the one (1) example found in John 4: 48 and John 20: 29.

In John 4: 48 a lack of seeing "signs and wonders" (which by specific instance includes seeing Christ in person) implies unbelief, but yet in John 20: 29

'belief' can exist without the specific sign or wonder of seeing Christ in person. Notably, Matthew 13: 58 is the explicitly stated converse of John 4: 48, which by the rules of formal, human reasoning would result in the humanly logical equivalence of unbelief and lack of seeing signs and wonders, yet John 20: 29 indicates that not only is it the case that such logical equivalences are not always assumed by Christ's Use of Logic, but more generally that a given conditional/procedural statement can have even explicit overlap with some parts of the formal rules of human reasoning without matching every rule of formal, human reasoning. Also, in John 2: 23, the inverse of John 4: 48 holds as well. Thus, a given instance of logic in the words attributed to Christ, while in one (1) aspect can be entirely separate from formal, human reasoning, may in all others be completely consistent with formal, human reasoning.

Moreover, strict rules of human reasoning do not govern the lives (actions or attitudes) of men, especially not from one person to another. The other forms, even those called "MAGR", are less clear in regard to the contrapositive, for if God does not do a particular thing, then man is still *not necessarily required* to act in a particular manner (unless one assumes a strictly deterministic worldview in which the concept of free will for mankind is not applicable). Also, the form called "GAMR" is such that if man does not do a particular thing, then God does *not always* necessarily required not to do something else, though in some cases he seems to say he will not do certain things if man does not act such as in item # 80 in which man may not "freely give", yet God has not bound Himself NOT to still freely give to that person who does not "freely give". Importantly, of the 240 statements involving "GAMR", only nineteen (19) are not considered to be false when compared to strict rules of human reasoning, though none of these involve definite statements in Christ's Words to affirm this finding. Lastly, the form called "GAGR" has instances in which the contrapositives are false (#160), but in other the contrapositive may be true (# 139). There are other forms in which a man's and God's actions and/or results are joined together, either implicitly or explicitly.

The most common thread, though not entirely, that runs through the instances in which the contrapositive of a conditional/procedural form attributed to Christ is considered "false" can be described as follows: Whenever the condition of a logical form is ALWAYS true, then the conclusion of the contrapositive must necessarily ALWAYS be false (by strict rules of human reasoning). Further, when this occurs along with the condition of the contrapositive being true, then by strict rules of human reasoning, the contrapositive will be false as an overall logical form.

The difficulty with this though from the perspective of the logical forms attributed to Christ is that, by strict rules of human reasoning, the logical form itself too would also be false (which would not be problematic from the strict rules of human reasoning) if the contrapositive were considered to be false. Thus, if the logical forms attributed to Christ are assumed always to be true (as overall logical forms/statements), then by the strict rules of human reasoning, whenever a condition is always true, the conclusion of the logical form must also be true as well. Additionally, since the condition of the original condition must always be true, the only other option for the contrapositive to be true is when both the condition of the contrapositive form is false so that the contrapositive is true vacuously.

Hence, the keys to this apparent discrepancy are that: a) the logical forms attributed to Christ are assumed to be true overall, which restricts the possibility, by the strict rules of human reasoning, of having a "true" condition with a "false" conclusion implying a "true" logical form overall and which is most definitely not true in strict rules of human reasoning; and b) there is no indication in the Word's attributed to Christ that a statement can be vacuously true. Notably, there are four (4) instances of the contrapositive of a logical form attributed to Christ—two of which are both attributed to Christ as being true First, Matthew 12:30 and Mark 9:40 are 'true' contrapositive statements (see # 116 and # 482 above). Second, here is also John 14: 21 and 24 (see # 1264 and # 1267). Third, there is John 5: 23 (see # 1055 and # 1480). There is one (1) other instance where a statement is "true", but the contrapositive is stated explicitly as being "false" (see John 20: 29 and John 4:48).

In John 4: 48 a lack of seeing "signs and wonders" (which by specific instance includes seeing Christ in person) implies unbelief, yet in John 20: 29 'belief' can exist without the specific sign or wonder of seeing Christ in person. Notably, Matthew 13: 58 is the explicitly stated converse of John 4: 48, which by the rules of formal, human reasoning would result in the humanly logical equivalence of unbelief and lack of seeing signs and wonders, yet John 20: 29 indicates that not only is it the case that such logical equivalences are not always assumed by Christ's Use of Logic, but more generally that a given conditional/procedural statement can have even explicit overlap with some parts of the formal rules of human reasoning without matching every rule of formal, human reasoning. Also, in John 2: 23, the inverse of John 4: 48 holds as well. Thus, a given instance of logic in the words attributed to Christ, while in one (1) aspect can be entirely separate from formal, human reasoning, may in all others be completely consistent with formal, human reasoning.

As such, no differencing is required to show definitively that Christ's Use of Logic did not assume that the formal human rule for contrapositives and statements being true or false together as a set would always hold, which lends additional credence to the possibility that the other instances above may (though not necessarily) be examples of this type of ‘exception’. Finally, this demonstrates explicitly that, in addition to the primary ‘exception’ for ‘double-negatives’, Christ’s Use of Logic is not a proper subset of the formal rules of human reasoning, though notably in more cases than not where explicitly stated, the contrapositive does hold in Christ’s Use of Logic.

E. Separate Conditions Implying the Same Conclusion: This is the most common form of all used in the words attributed to Christ, but is distinguished from the form where “If A implies B (with) since C implies B” since this latter form only provides an explanatory note to the original condition. Additional conditions implying the same conclusion has no inherent concerns in regard to formal human reasoning. In this regard, consider forms # 96 and # 479 as examples.

F. Partial Conditions Implying Opposite Conclusions: While not occurring frequently in the conditional/procedural forms in Christ’s Words, this logical structure in Christ’s Words is seen in logical structure # 22 (see items # 521 and #890 for a good example of this). This logical structure does not imply a violation of the formal rules of human reasoning, though it might seem to some casual observers that there is a sense of violation in this logical structure, but it is important since this logical structure points to the criticality of noting all conditions in order to reach the same conclusion as an original statement does. In the case of # 521 and #890, desire is not sufficient alone to achieve a “desired” end always (See #890), but when “belief” and “prayer” are added to the “desire”, then the “desired” end is promised to be reached.

Chapter Four

General Conclusions

Given that the first main goal of this analysis has been to document all of the conditional/procedural forms in the Words attributed to Christ as are contained in Appendix A and having derived the logical categories into which the conditional/procedural forms can be contained (see Chapter 3) as per the second main goal of this study, the focus of this chapter is upon addressing the logical implications and general conclusions relative to the third goal of this study, which was to compare the conditional/procedural forms to the formal rules of human logic.

GENERAL CATEGORIZATIONS OF 'EXCEPTIONS' NOTED FROM THE CHRIST'S CONDITIONAL/PROCEDURAL FORMS IN COMPARISON TO THE FORMAL RULES OF HUMAN LOGIC

Aside from the four (4) types of noted 'Exceptions', Christ's Words do not generally differ with the formal rules of human reasoning. Each of these four (4) types of 'Exceptions' are though quite important since the nature of these issues are foundational to the very nature of the formal rules of human reasoning, where the fourth type of 'Exception' is noted as an occurrence specific to the use of the King James Version of the Holy Bible for this study. As noted, the first three (3) types of 'Exceptions' are intimately linked in terms of things being permitted to have multiple 'truth values' (that is, for both A and NOT A to be true at the same time). This is very significant since such negations would technically challenge all of the formal rules of human reasoning involving negatives, which also includes negations of both the condition and conclusion for contrapositives {see Exception 2)}, though notably no such challenges involving negations using only the words attributed to Christ in the Holy Bible have been found from this analysis. Moreover, for the first type of 'Exception', which involves explicit indication of such multiple 'truth values', each of the six (6) instances of this type implies according to the formal rules of human reasoning that the statements so affected would be considered 'false', which is in conflict with the assumption of this study that all of the conditional/procedural statements attributed to Christ are true. The issues with these 'Exceptions' though are not limited to strictly logically aspects. By way of contrast with the first three (3) types of 'Exceptions', the second instance of the fourth type of 'Exception' noted from this analysis is also significant not so much logically as theologically in that the main issue in this fourth type of 'Exception' is not specific to multiple 'truth values' as with the other three (3) types of 'Exceptions', but rather to nature of the relationship between "(God the Father's) commandments" and "(a person) know(ing) God . . . and Jesus Christ".

A crucial observation is that there are no formal rules of human reasoning that would not technically be impacted by these 'Exceptions'. Very notably though, there is only one (1) instance of another formal rule of human reasoning being impacted negatively (which is the rule of 'A NOT implying NOT A'—see # 1016 in the discussion on 'violations' below).

Thus, while there would theoretically or potentially be very wide-spread implications from the (4) types of 'Exceptions' if using only the standard of comparing Christ Use of logic to formal rules of human reasoning, the actual or realized 'Exceptions' are limited to a total of nine (9) 'Exceptions'. Specifically, these nine (9) 'Exceptions' can be categorized according to the formal rule of reasoning via three (3) types of exceptions for which there is a exception(s) or according to the central, underlying theological or technical issue at question:

EXCEPTIONS CATEGORIZED BY DIFFERENCES WITH FORMAL, HUMAN REASONING RULES

A) There are nine (9) involving exceptions of the double-negative rule—see Exceptions in 1)
B) One (1) is specific to a exception of the contrapositive form—see Exception 2) ; and
C) One (1) other is specific to an exception of the dual inclusion rule for equivalence relationships.
D) Two (2) instances of lack of transitivity for equivalence relationships.

This first categorization has already been explored in the above chapters. So, the remainder of this section will focus on a second type of categorization that specifies the underlying theological or technical issues associated with nine (9) of these 'Exceptions' since 1.f) is a direct result from the use of the King James Version, as is 'Exception 4).

In particular, this other way of categorizing these nine (9) significant 'Exceptions' yields the following:

Significant Exceptions Categorized by Underlying Theological or Technical Issues

i) The Trinity
ii) The nature of human belief in Christ
iii) The nature of "good(ness)" for believers and unbelievers
iv) The nature of the relationship between "(God the Father's) commandments" and "(a person) know(ing) God . . . and Jesus Christ"
v) The promise of Divine Protection for the disciples

EXCEPTIONS DIRECTLY RELATED TO THE TRINITY

It is highly noteworthy that the relationships within the Trinity (including God the Father, Jesus Christ, and the Holy Spirit) is the most common underlying factor from which these 'Exceptions' arise as this factor is involved directly in six (6) {see Exceptions 1.b), 1.c), 1.d), 1.e), 1.g), and Exception 3)} of the eight (8) 'Exceptions'. In contrast to where there are differences noted between the members of the Trinity, there are also instances involving dual inclusions (see # 1616, # 1621, and # 1624) that all state "Christ in God the Father" and "God the Father is in Christ". This is significant since formal human reasoning would thus equate Christ and God the Father and yet all the more reinforces the importance of the instances where specific differences are evidenced within the Trinity.

This is so notable since both the logician/scientist and the Bible scholar adhering to a Trinitarian interpretation of the Holy Bible (which is the majority viewpoint of modern Christianity) accept that the Trinitarian view of one (1) "God-head" with three (3) distinct persons, where one (1) of those persons, specifically God the Father is "greater than" the other two (2) persons at least in the sense of both "proceeding from" the God the Father, defies a fully complete and consistent description with the words available to mankind. Indeed, Christian Bible scholars adhering to a Trinitarian perspective describe the human inability to describe any of these persons fully as "transcendence" (which means effectively 'that which is beyond the capacity of mankind to know fully'), and the logician/scientist describes the Trinitarian viewpoint as either flatly illogical or at best an erroneously worded human construct in the metaphysical sense. The view that the Trinity is strictly illogical results from appealing to a exception of the law of double negations of an object not equaling the original object given that such equality can only be preserved when separately considering the Divine traits from the functions of the members of the Trinity. Notably, such concerns have been the most common theme of debates (and resulting identification of 'heresies') within Christianity regarding doctrine and theology, with most centering on interpretations of the Biblical texts regarding statements specific to Christ's Humanity and Deity. The view of the Trinity as an erroneously worded human construct results from a number of factors, including de-constructive interpretations and those, such as the metaphysicists, which accept the existence of a reality or realities beyond the current or possibly potential knowledge of mankind. There is overlap between these two (2) generally contrasting views—namely, that of limitations in terms of human language (though expressed by the Bible scholar from the perspective of inherent, human limitations relative to God, which is a different viewpoint of a similar effect from the de-constructivists who would point to an inherent limitation in terms of all human language, including the written word, to convey absolute meanings, though notably the stricter forms of de-constructivism would disagree with the very premise of this book, which is that there are definite, literal meanings to the words attributed to a person, specifically Christ in this case).

However, the critical point in this matter for the purposes of this study is that none of the statements in question regarding the Trinity affect the assumption noted previously regarding Jesus Christ as having the same nature and attributes as God the Father. The central question posed by these 'exceptions' is rather one of the hierarchical authority of Jesus Christ in relation to God the Father since the vital point of difference here is that Christ is to have said that ". . . (God the) Father is greater than all" and specifically ". . . greater than I (Christ)." This assumes that, as noted previously, the instance in which Christ is to have said, "Why callest thou me good; there is none good, but God?" is intended within the context of that passage in a strictly rhetorical manner and as such is not in conflict with the other statements attributed to Christ.

It is well beyond the scope of the analysis to discuss all of the details regarding a Trinitarian interpretation of the Holy Bible. So, the main point of the above discussion is that such 'exceptions' as the four (4) noted related to the Trinity are not unexpected and also are not fully explainable in human terms.

The other two (2) main 'Exceptions' are directly connected to the nature of man in regard to 'belief' in Jesus Christ as the Son of God and to the capacity of man to possess the godly quality of 'goodness'. These are Exception 2) and Exception 1.a)

EXCEPTIONS DIRECTLY RELATED TO THE NATURE OF HUMAN BELIEF IN CHRIST

As regarding Exception 2), which involves mankind's 'belief' in Jesus Christ as the Son of God, while there is clearly as aspect of Trinitarian thought here, the focus of the pertinent statements here is the role mankind to such truths both with and without miraculous "signs and wonders" as tangible confirmation of such. The term, 'mankind', is indicated here since the issues and responses for 'belief' are not specific to a particular individual since the word 'ye' is used (which is the plural form of the second person) in # 1045 (see John 4: 48), since 'they' is used in # 1396 (see John 20: 29) , and since the message of Christ's Gospel had already been opened technically, though not broadly at that point, to the Gentiles by the announcement to the Magi (who were Gentiles) at Christ's Birth.

Even closer examination of the statements involved here (see # 1045 and # 1396 in Appendix A) reveals that essentially the relationship between the set of people who saw "signs and wonders" (call this set "S&W") and those people who had "belief in Christ (as the Son of God)" (call this set "B in C") based on the words attributed to Christ in the Holy Bible is such that the "S&W" and "B in C" are intersecting, but neither is contained inside the other; however, this relationship prevents the contrapositive of # 1045 from also being true. Hence, the explanation for "belief" or non-belief is not specific to the "signs and wonders", but rather to the personal 'willingness' (assuming free will on the part of the one who ultimately believes in Christ) or upon a factor belonging to God (assuming "Sovereign election"). Importantly, neither of the primary passages here (John 4: 48 or John 20: 29) mention any sense of the later possibility, and "the elect" is only mentioned three (3) times in the conditional/procedural statements attributed to Christ in the Holy Bible , all of which occur in Matthew 24 (see # 303, # 304, and # 313 in Appendix A), for which the context is not generalizable beyond the times associated with the "sign of thy coming, and of the end of the world (Matthew 24: 3). Also, see John 6: 39 and 40 (see # 1087 and # 1089 in Appendix A) for a potentially implied connection between "those given (Christ) by God the Father" and those who "seeth the Son, and believeth on him" since these groups are the subjects of statements explicitly stating "(God the) Father's will" which promise that both groups will be "raise(d) up . . . on the last day." This is potentially significant since such would imply by the rules of formal human reasoning that the two (2) groups (those "given (Christ)" and those who "seeth him, and believeth on him") are the same group. However, there is no other statement attributed to Christ by which such a connection is explicitly stated. Therefore, based solely on the words attributed to Christ in the Holy Bible, the only firm conclusion here is that "belief on Christ (as the Son of God)" must occur in those persons who are to have "eternal life", but the mechanism for how such belief occurs *precisely* in all persons is not explicitly identified (*). Indeed, the use of literal interpretations for the words attributed to Christ in the Holy Bible yields that both God's Act of "giving (persons to Christ)" and the act of a person expressing "belief in Christ (as the Son of God)" are assumed in the logic employed by Christ in the Holy Bible for those who obtain "eternal life".

(*)—See # 1094, # 111, # 1125, and # 1138 for instances specific to Jewish persons in which statements attributed to Christ in the Holy Bible imply directly that "belief in (God) the Father" would result in "belief (in Christ), though none of these statements reference "the elect" in any manner. Moreover, such statements cannot be generalized since Old Testament teaching explicitly identifies the Jewish people as "God's chosen people" through promises made to Abraham, though even for the Jewish people it is clear from Christ's statements (such as # 1069 in Appendix A) that not all Jewish people at the time of Christ believed in Christ as the Son of God and hence did not truly believe in God the Father, albeit by physical birth part of "God's chosen people".

EXCEPTIONS DIRECTLY RELATED TO HUMAN "GOOD(NESS)"

As regarding Exception 1.a), which involves the capacity of a certain type of man to be described as "good" in one place, but in another place for the words attributed to Christ in the Holy Bible to indicate that "none is good, but God", the reason for the exception is that a specific individual of a group cannot be both "good" and "not good" at the same time, where notably the same original Greek word for "good" is used in both places as an instance of violating the rule for double-negatives. The context of Luke 6: 45a (see # 1093) only tells one that a "good" man is part of those groups for whom "good fruit" is produced and who have "good treasure" in their "heart". The context of Matthew 19: 17 (see # 1487) leads one to assume that Christ is using this statement to reinforce his own personal quality of "goodness" as the Son of God, with Christ notably not denying that he is "good" as the young inquirer addresses Him as "Good Master". In so doing, an element of the Trinitarian viewpoint is introduced again, but this is not the primary issue at hand in Exception 1.a) since the exception occurs because humans (not Christ) are described in Luke 6: 45 as "good" in the same sense as Christ is to have said in an unqualified manner in Luke 6: 45 that only God is "good". Further, Matthew 5: 13 and Mark 9: 50 (see # 1493 and # 492 in Appendix A) would indicate by application of the transitive rule of equivalences that disciples of Christ are "salt" and as such are "good", which is the same word as above for 'good'. However, since the transitive rule for equivalences does not always hold in the statements attributed to Christ and since Christ did not explicitly use such 'transitive' logic for equivalences nor did he state that the 'disciples' are/were 'good', this specific aspect for 'goodness' of the disciples of Christ does not need to be pursued further in this regard.

There are no other passages of concern in this regard in the words attributed to Christ in the Holy Bible, though this basic passage is effectively repeated in multiple books of the Gospels that directly address these concepts. So, there are no explicit indications in the words attributed to Christ in the Holy Bible that directly state the implications of "good(ness)" in relation to groups or individuals (including most importantly for believers or unbelievers). Additionally, the words attributed to Christ indicate directly in # 17 in Appendix A that adherence to the law or external performance to the 'letter of the law' is not an adequate measure of "good(ness)" for believers in Christ, and in # 151 and # 152, the "heart" is explicitly identified as the source for both "corrupt" words and for "good" fruit. Thus, the central question here is whether or not "corrupt" and "evil" *when applied to humans* are not literally opposites of "good" (that is "NOT Good does not equal "corrupt" or "evil" when applied to humans, either before or after "belief in Christ" for believers or in general for unbelievers). While Luke 6: 43 and Luke 6: 45b both indicate clearly that not only will "corrupt" fruit or "evil" result from an "evil" heart, but that "corrupt(ness)" cannot bring forth "good(ness)", there are numerous indications in the words attributed to Christ, even for the disciples, which demonstrate that Christ did not assume those who "believe" in Him would be entirely "good" in the same sense as God is "good" (for example, see # 469 for an instance in which Christ said to Peter, who was one of the three (3) closest disciples of Christ, that Peter "savourest not the things of God" and that effectively Satan was working through Peter at that particular time.

Hence, if restricted to literal meanings only and to the definitions of "corrupt", "evil", and "good" only "good(ness)" completely excludes any "corrupt(ness)", and vice-versa, Exception 1.a) points not to a theological issue as with the other instances, but a result produced by the over-arching assumption of literal interpretations as used exclusively in this book and the associated effect from such literal interpretations upon matching "yes/no"-type definitions of human qualities to persons (both believers and unbelievers) who are composites of both "good(ness)" and "corrupt(ness)" to varying degrees. It may be argued that the 'intent' in Luke 6: 45 is that the 'primary desire' or 'core' of a person's self may be "good" and so qualify to be called "good" if that person is a believer in Christ, and this is consistent with the totality of Christ's words and that of the Holy Bible regarding believers in Christ. If the person is not a believer in Christ, then no "goodness" can be assumed in the sense of God's "Goodness", and the binary view of "goodness" or "no goodness" applies since the "goodness" must come from God initially through belief in Christ as Christ stated repeatedly that being in and having a right relationship with God, who is the one completely "good" entity is directly related to belief in Christ. However, within the assumptions of this analysis, which are required for its very execution since formal human reasoning does not permit such 'composite' definitions, such a 'majority' view is not applicable to the contexts of the passages in question from Luke 6: 45b and Matthew 19: 17. It is ultimately vital to note that, under different assumptions regarding "good(ness)" as a 'primary desire' or 'central perspective' of a human believer, which is contrast to complete "good(ness)" or complete "corrupt(ness)", it is clear from the above analysis that no exception would be necessary under such an assumption, with all other factors being equal as they would be with the fixed passages in view here.

AN EXCEPTION DIRECTLY RELATED TO "KNOWING GOD AND JESUS CHRIST" AND "(GOD THE FATHER'S) COMMANDMENTS"

While the first instance of the fourth type of 'Exception' can be attributed to use of the King James Version of the Holy Bible for this study, the second instance of this type, in which 'transitivity of equivalence relationships' does not always hold via literal interpretations, requires further attention. Specifically, from # 1458 and # 1487 in Appendix A, an instance of this lack of transitivity for equivalences is found comparing John 17: 3 and John 12: 50. Although no aid is found from the words attributed to Christ regarding "eternal life", which has the same Greek origin as "life everlasting", # 205 is another supporting statement much like # 1487 found in Matthew 19: 17 in which

Christ is to have said to the young ruler, "if thou will enter into life, (then) . . . keep the commandments . . .". Such "keep(ing) of his (God the Father's) commandments" is importantly summarized in # 1441 and # 1442 from Matthew 22: 36-340 as being directly conditioned by two (2) types of "love"—specifically, (of) "The Lord thy God" in an total sense (that is, with all human capacities) and "thy neighbor as thyself". This same principle also applies in significant detail to "keeping of (Christ's) commandments/sayings" and "love (of Christ)" from John 14: 21-24 (see # 1264 through # 1267 in Appendix A).

Thus, in order to understand keeping "His (God and Christ's) commandments", the words accorded to Christ turn one's attention toward "love (of God and Christ)" in the sense of self-giving desire and sacrificial acts as is directly implied by the use of *'agape'* for love" in these passages. Hence, selflessness both in attitude and action is implicative of such 'love" and accordingly from above of "keeping (God and Christ's) commandments.

One must though ask though how does such selflessness equate or even relate to "know(ing) God" where such? To address this vital question, one must first explore what the word "know" means in the Gospel According to John. The Greek word for "know" in John's Gospel has both the sense of 'process or coming to learn more' and 'understand', which are the two (2) non-idiomatic meanings of this word for "know" in the New Testament. In this light, the question here then becomes: does 'selflessness' equate to or even directly relate to either such a 'process or coming to learn more' or to such 'understanding'. Since it is assumed from the words attributed to Christ that (God the Father) "is greater than all" and that "I (Christ) and the Father are one" as discussed above in the section regarding 'Exceptions' specific to the Trinity, it is assumed that Christ did not mean to imply that "know(ing) God and Jesus Christ" means 'understanding' in the sense of 'complete knowledge', which is the case independent of one's epistemological viewpoint regarding human knowledge and its acquisition. However, there is no explicit indication in the words attributed to Christ that equates any words or even concepts involving 'selflessness' to 'coming to learn more' about either God the Father or Christ'.

There is though a line of implications beginning from # 1132 in Appendix A to # 1410 and culminating in # 1248 and # 1249 that can permit direct implication of "continu(ing) in my (Christ's) word" to "know(ing) (Christ)" and hence according to # 1248 and # 1249 "know(ing) (God the Father)". This three-part linkage is hinged on # 1410 in which Christ is to have said "I am the truth . . ." since the first half of the conclusion in # 1132 is "know the truth". This is not meant to imply that "(Christ's) words" are all of (God the Father's) commandments, albeit Christ is accorded as stating in several places, such as in # 1129 from Appendix A, that Christ only spoke what He heard from (God the Father), which is presumed then to be a literal subset of the complete (God the Father's) commandments, since Christ's words do not encompass explicitly all of the commandments given by God the Father at all times, particularly not including all of the commandments of God the Father from the Old Testament. This issue of Christ's words being a proper subset of the words from God the Father is resolved though by # 1492 in Appendix A which states that "thy (God the Father's) word is truth". There are caveats though to this linkage. Specifically, the original equivalence relationship here was not keeping or continuing in "(God the Father's) commandments", but rather just "(God the Father's) commandments". Also, it must be noted that none of the implicative linkages via connecting conditional statements results in a direct equating of the "(God the Father's) commandments" and "know(ing) (God the Father) and Jesus Christ", albeit as seen previously such transitive equivalence implications are not in conflict with the logic employed by Christ, but rather that no single conditional or procedural statement (including no 'one-way' inclusion) is logically equivalent to an actual equivalence relationship.

EXCEPTIONS REGARDING DIVINE PROTECTION OF THE DISCIPLES

There are two (2) instances of this type. See Exceptions 1.h) and 1.i).

First, from the statement, ". . . nothing shall by any means hurt you." in Luke 10: 19, which is in conflict with Matthew 24: 9 in which Christ is to have said that when Christ's statements regarding the end of time are fulfilled, then the disciples would be "delivered up to be afflicted", and they would be "kill(ed)". Mark 16: 18 makes a similar promise of Divine Protection as is found in Luke 10: 19, but there such Protection is conditioned specifically to "tread(ing) on serpents" and "drink(ing) (anything of a poisonous nature)", whereas there is notably no condition stated or implied in the promise for Divine Protection in Luke 10: 19. Since there is no condition other than "nothing" for Luke 10: 19, including no timing or procedural aspects, there is no mitigation seen from a literal interpretation of these words as any such mitigation would need to in some manner address the possibility that the disciples would not be "harm(ed)" (which includes any physical harm up to and including being "killed)") up to some specific point (i.e., when they would "delivered up to be afflicted"). (See # 1627 and # 290 in Appendix A). Also, see below for how Luke 10: 19 also involves a 'violation'.

Second, there is a very similar 'exception' to in Luke 21: 16, in which ". . . some of you shall they (parents, brethren, kinsfolks, and friends) cause to be put to death" that is realized since Luke 21: 18 also states that, "But there shall not an hair of your head perish". The compounding factors in this instance though are that the context is the same in both statements since they occur in the same chapter of Luke and that the condition of "some of you (the disciples) shall they . . . put to death" precedes the conclusion of "shall not an hair of your (the disciples') head perish . . .". (See # 1628 and # 1629 from Appendix A). These factors do not permit mitigation of this 'exception' within the bounds of literal interpretations only as applied in this study. Also, see below for how # 1628 involves a 'violation' as well.

ASSUMPTIONS REGARDING EXCEPTIONS

The instances where contextual information for instances that initially appeared to be 'exceptions' results in no actual 'exception' occurring are not comprehensively listed as 'potential exceptions' nor are any other 'exceptions' that only appear to be so initially, yet which can be explained by other means. An example of the criticality of contextual information is found in a set of statements (see # 39, # 43 and # 927 from Matthew 5: 44 and Luke 19: 27, respectively) that appear to be 'exceptions' since # 927 can appear literally to be at odds with "lov(ing) your enemies" as given in # 39 and # 43, but which are considered not to involve an 'exception' since the passage in Luke 19: 27 is literally applied to the timeframe of "till I come" (which in the context of that passage is indicated as the 'time of judgment when Christ returns'). As a further example regarding 'potential exceptions', no instances are analyzed in detail for which Christ is to have stated a particular thing and another person is separately to have said or indicated something that would be at odds with such as the words attributed to Christ are only compared to other words accorded to Him. Lastly in regard to 'potential exceptions', since strictly declarative (non-procedural and non-conditional via direct or inferred means) statements attributed to Christ are not considered in this study, instances such as John 20: 27 (that is strictly declarative) and John 20: 17 which involves an 'exception' of the first type for 'double-negatives' are not considered.

GENERAL CONCLUSIONS REGARDING VIOLATIONS OF FORMAL HUMAN REASONING

Approximately 82% {which is forty (42)} of the fifty-one (51) total instances of 'violations' of formal human reasoning from comparison using literal interpretations only of the words attributed to Christ involve a definitional statement pertaining to traits of Christ and/or God the Father, and as result issues specific to physical impossibilities related to the Trinity are the predominant reasons for the noted 'violations'. There are nine (9) other 'violations', and these pertain to a range of theological issues and result from one of four (4) situations:

a) unconfirmable statements specific to different aspects of 'life after death' (see # 1058, # 1079, # 1101, # 1119, # 1184, and # 1205—note: not all statements involving 'everlasting life', 'life everlasting', 'eternal life', or 'life eternal' are represented here since the same precise 'violation' occurs in all places where such is used);
b) conflict with a known rule of formal human reasoning, such as 'A does not imply NOT A' (see # 1628); and also
c) historically verifiable evidence that conflicts in the literal sense with a given statement attributed to Christ (see # 1627); and
d) physical impossibilities (see # 1016, # 1493, # 1494, and # 1511).

Thus, the noted 'violations' most frequently result from physical impossibilities in which Christ and/or God the Father are equated to physical objects, such as "bread", "light"; or to each other using the verbs; to the disciples or 11 Apostles: 'am', 'are', 'is', 'be', or in one (1) case, 'have been'. (See Appendix A for #1399-1412, # 1421, # 1432-1438, # 1440, # 1443, # 1445-1446, # 1468-1471, # 1473-1476, # 1402, # 1493-1495, # 1511, # 1514, # 1023 with # 1210, and # 1095 with # 1249.)

Of the fifty-one (51) total 'violations', forty-four (44) are strictly 'violations' and as such are not associated with any 'exception'. Thus, only seven (7) of the total 'violations' also involve 'exceptions'. All, but one (3), of the 'non-exception violations' are found in: # 1271 with # 1412; # 1095 with # 1249; # 1475; and # 1023 with #1210—{specifically, see 'Exceptions 1.c), 1.d), 1.e), and 1.g), respectively}. Exceptions 1.c) and 1.d) requires 'one(ness)' of members of the Trinity that are stated explicitly as being different altogether in terms of hierarchical authority (i.e., God the Father is "greater than" all, including Christ}. Exceptions 1.e) and 1.g) explicitly indicate in the condition of the statement that a certain trait belongs to Christ, but then in the conclusion states the exact opposite, which according to the formal rules of human reasoning makes the entire statement 'false'. Hence, these first four (4) instances all involve issues specific to the Trinity as examined earlier in this chapter.

These 'violations' related to the Trinity also result in logical differences with the syllogistic forms from formal, human reasoning in that the specific condition of Christ's being 'one with God the Father' and also that Christ is included in those who are not 'greater than God the Father' both does and does not permit the general application of syllogistic inferencing. There are, however, two (2) examples of explicit syllogisms, one of which involves Christ's and God the Father's will (see # 1086 and # 1087 in John 6: 38-40) and the other of which involves the anyone, including the disciples, not being able to bear fruit without abiding in Christ (see #1283 and # 1284 in John 15:4). There is also at least one (1) 'implicit' syllogistic form (see # 1110 and # 1111 in Luke 24: 44-47), where "all" being "fulfilled" results in the fulfillment of Christ's gospel being preached to the rest of the world (outside of Israel and the very few non-Jewish, non-Samaritan cities that Christ visited during his earthly ministry), where the implicit condition to complete syllogism is provided by the context of the crucifixion and resurrection for these statements. Thus, while no syllogistic forms involve 'exceptions', sufficient statements can be connected together across the statements attributed to Christ to demonstrate that no generalized use of syllogistic reasoning can be assumed here.

In addition to the four (4) 'non-exception violations' involving the Trinity, there is also a second set of three (3) 'non-exception violations', which do not pertain to the Trinity directly. The first two (2) of this type of 'violations' involve promises related to Divine Protection of the disciples. First, from the statement, ". . . nothing shall by any means hurt you." in Luke 10: 19 (see #1627 in Appendix A), which later is not true by a strictly literal interpretation since history records that all of the disciples, except for Judas and possibly John were martyred. Second, from Luke 21: 16 (# 1628 in Appendix A), in which part of the condition of this statement includes ". . . some of you shall they (parents, brethren, kinsfolks, and friends) cause to be put to death" results in a 'violation' since the conclusion of # 1628 indicates, "But there shall not an hair of your (the disciples') head perish" assuming that the other parts of the conditions for # 1628 are 'true'. Given that this 'violation' occurs within the same conditional statement, no contextual or timing/procedural-related mitigation to this 'violation' are deemed appropriate. From the logical perspective of comparison with formal human reasoning, this second 'violation' occurs since 'A cannot imply NOT A'. The third case of a 'non-exception violation' is noted from John 3: 3 (#1016), in which the condition of "born again" is assumed in order for one to see the kingdom of God since as noted by Nicodemus, such is not physically possible if a strictly literal interpretation is taken, though this is not an 'exception' to any other statements attributed to Christ and is explained in John 3: 5-8.

Of the seven (7) cases where a 'violation' also involves an 'exception', there are three (3) cases (see # 1440, #1511, and # 1402 with # 1494 in

Appendix A) which pertain to the nature of the members of the Trinity, and are as follows:

First, # 1440 involves a 'violation' of formal human reasoning in which Christ is to have said, "All things are possible with God . . .". See Luke 18: 27 and also "what is impossible with men is possible with God" in Mark 10: 27. The 'violation' of formal human reasoning from # 1440 can be stated more generally as: Could God create an object which He could not affect or could He not create such an object? Importantly, any arguments on either side of the debate in this particular paradox involving the concepts of 'omnipotence' or 'infinity' must not be impinged beyond 'all things' if a literal rendering of the words is applied, even considering the context of this statement, as is assumed in this study. Once these concepts are so specified via literal interpretations solely, God would be 'limited' in one way or another, though it must emphasized that such a 'limitation' is derived strictly from comparison of this statement to the formal rules of human reasoning since no immovable object and irresistible force can co-exist according to formal human reasoning.

Matthew 17: 20b, where Christ is to have said to the disciples that ". . . nothing shall be impossible unto you", is seen as a specific application of the larger issues from this classic paradox. Matthew 17: 20b does not though involve an 'exception' since the condition for this statement in Matthew 17: 20b is "faith (in God)" and since the context for Matthew 17: 20b is primarily prayer to God and healing of others from demonic possession. There is one (1) other statement that must be addressed in this regard since a literal rendering of John 15: 5 could also be viewed by some as an 'exception' to Matthew 17: 20b since John 15: 5 states, ". . . without me (Christ), ye (the disciples) can do nothing." However, this statement reinforces again the conditional character of most of the statements attributed to Christ since Christ did not state unconditionally that "ye (the disciples) can do nothing", but rather "without me (Christ)" is the condition under which this statement is considered to be true. Moreover, John 15: 5 is effectively the 'inverse' of Matthew 17: 20b, which as has been seen previously is generally 'true' in the words attributed to Christ, though such is not required by formal rules of human reasoning nor absolutely in the words attributed to Christ. In this light, Matthew 17: 20b is not in conflict with John 15: 5.

The second special case from # 1511 involves a 'violation' of formal human reasoning in which the disciples are said to be "not of this world". This is a 'violation' since using literal interpretations only no human can "not be of this world", where "world" implies the "cosmos" or the material world. Notably though, # 1514 from Appendix A, in which Christ is to have said, "I am not of this world" does not involve an 'exception' since neither # 1511 or # 1514 are a definitional statement, though it does involve a third special case of this type of 'violation' for the same reason as for # 1511 in considering the humanity of Jesus Christ.

An even more specialized third case is found in # 1402 and # 1494 from Appendix A, where the equivalence of "I (Jesus Christ)" to "the light of the world" in John 8: 12 is not the same literally in comparison to "Ye (the disciples following Christ)" from Matthew 5: 13, albeit all the Greek words are the same for "the light of the world". This equivalence is not the same regardless of which group of 'disciples' (i.e., those there at that time, all present and future disciples, or those who meet the conditions in verses 3-12 of Matthew 5) as Jesus Christ is not the same entity as any group of disciples. There is a partial mitigation of this issue found in a dual inclusion from John 6: 54 and 56 (see #1101 and # 1103 as well as # 1620 and # 1622) the Apostles, from John 17: 23, 26 (see # 1495 and # 1496), in which a 'kenotic' sense of equivalence between Christ and the disciples of Christ is implied. For # 1101 and # 1103 the same condition of "eatheth my blood and drinketh my blood" precedes both "eternal life" and "dwelleth in me, and I in them" so that this dual inclusion applies to all who have eternal life, which is the same as all disciples, not just the Apostles. This mitigation is also not different across any of the major translations of the Holy Bible. Thus, this is only a 'violation' of formal human reasoning and not an 'exception' since these statements lead to no conflicts from strictly an internal comparison of the words attributed to Christ, but the physical reality of Christ not being one (1) with any group of disciples is a strict violation of the formal rules of human reasoning since Christ as an individual cannot be literally the same as any other person(s).

ASSUMPTION SPECIFIC TO VIOLATIONS

None of the statements attributed to Christ that contain instances (though most do not) in which the condition has no directly tangible or strictly rational (via the strict rules of human reasoning) impingement on the conclusion (such as when Jesus says in John 9: 41, ". . . if ye were blind, then . . . ye should have no sin . . .") are considered to be 'violations'. In formal human reasoning, these types of statements would be always be true overall using this book's key assumption of literal meanings unless the case is considered of a person who is both "blind (literally) and has "no sin".

ADDITIONAL GENERAL OBSERVATIONS

More generally beyond the assumption for 'violations' above, it should be noted that these general conclusions do not consider those instances in which

Christ's Words would strictly speaking form what logicians would call "non-sequitirs" For example, see form # 169 from Matthew 17: 20 in this study ("If you have 'faith' as a mustard seed, then you shall say to this mountain remove hence to yonder place and it shall be removed."). This is an example of Christ "mixing" a procedural form (if, then) with a conclusion that in strict rule of human reasoning would be infeasible, if not illogical. No logical equivalences are defied in these forms, but in strictly human terms and when taking only a strictly literal view of the words attributed to Christ, there is no inherent reason why the conclusion should follow from the condition. As stated in the assumptions of this study, no attempt is made here to address interpretations other than those that take a "literal" view of the words attributed to Christ. Many theological texts are written that address this matter in depth, including hermeneutics that aid in determining "symbolic" versus "literal" interpretations, though no studies are known that specifically attempt to compare such statements to formal logic. Thus, this is a matter that must be noted from an objective standpoint, but which is beyond the scope of this analysis. There are four (4) other examples of 'non-sequitirs'—see John 9: 41 (# 1156); John 6: 35 (# 1082); John 4: 14 (# 1037); and John 11: 9 (see # 1184).

The question of the different words for "if" in the Bible is an interesting one. There are eight (8) different words for "if" in the OT, and 7 different words for "if" in the NT (actually, 8 different words, but one of these is a derivative of another Greek word). In the Holy Bible, there are, however, only two different words with actually one Greek word (Strong's No. 1487), "ei", {which has the sense of how we would use "if" or "although"} being the most common by far (actually there are only 4 instances of the other Greek word for "if" in the Holy Bible, (Strong's No. 2579), "kan", (which has the sense of how we would use "even if"). So, aside from some possible meanings of "although" in our wording and a very few "even if" phrases, in the Holy Bible, the word for "if" in the words of Christ (using the KJV, which I have assumed in my research) seem to have very much the same sense as we would give to the word "if" generally.

GENERAL SUMMARY

From this evidence and more generally for those few equivalent conditional/procedural forms in Christ's Use of Conditional/Procedural Forms, we cannot conclude that the same rules of human reasoning for Conditional/Procedural equivalences (including contrapositives) *necessarily* apply *always* to Christ's use of Conditional/Procedural Forms or vice-versa. In a more positive manner, it can be said that the words attributed to Christ in the Holy Bible are very generally consistent with the logical forms and logical results associated with the formal rules of human reasoning. For example, the two (2) examples where Christ employed a statement and its contrapositive are reasonable, yet not conclusive, evidence that Conditional/Procedural equivalences are treated in the same manner in Christ's Use of Conditional/Procedural Forms as in the strict rules of formal human reasoning.

A strict note of caution in over-generalizing this result is that, while the strict rules of human reasoning are well defined, there is no body of "rules" to suggest the way in which Christ's Use of Conditional/Procedural Forms must operate. Thus, in the case of strict human reasoning, one can use deductive means to inference, yet with Christ's Use of Logic, one can only employ patterns for inferences from His Use of Conditional/Procedural Forms (though it has been noted that Christ's Use of logical forms primarily employ implications from general conditions, not individual or only exemplary conditions).

Specifically, since there is no formal set of rules dictating the axioms and logical patterns of Christ's Words, this means that, when the basis for comparison are Christ's Words, some instances in Christ's Words found not to be consistent with a formal rule(s) of human logic, can only be stated as a "potential" difference (for example, when the contrapositive of a "GAMR" form is thought to be false by rules of human reasoning, but do not have specific evidence of either falseness or truth in Christ's Use of logic). On the other hand, when the basis of comparison is a formal rule for human logic, instances of Christ's Use of logic that are not consistent with strict rules of human reasoning can be stated with a higher degree of deductively logical certainty as 'actual differences' with respect to the basis of formal, human logic (such as when a conditional/procedural form in Christ's Words has a contrapositive form that is not true by rules of strict human reasoning, where a condition, P, implies NOT P, or where dual inclusion cannot be assumed from an equivalence relationship).

More generally, there is no strictly deductive basis that can be applied to the 'exceptions' since the 'exceptions' are limited strictly to internal comparisons of statements attributed to Christ. The 'violations' can be stated with a higher degree of logical certainty from a formally deductive approach as is applied with the strict rules of human reasoning since literal interpretations of these words attributed to Christ and strict application of the formal rules of human reasoning permit a much greater level of deductive, axiomatic, and rigorous basis. This higher degree of logical certainty is not though entire or absolute since such ultimately would require that the axiomatic and logical structure of the words attributed to Christ be known fully in a manner comparable to formal human reasoning, which is not the case since there is no formal documentation elsewhere or indication from the words attributed to Christ as to the exact nature of the logic He employed. Furthermore, the documented instances of internal differences (called 'exceptions' in this study) are additional evidence that no strictly deductive logical basis for the words attributed to Christ can be applied, although in the main this analysis indicates that whatever the logical basis employed by Christ it primarily has significant similarities to formal human reasoning.

In formal terms the strict rules of formal human reasoning and Christ Conditional/Procedural Forms can only have five (5) possible interactions (no intersection, both sets are equal, one set is contained in the other properly (2 ways), or the intersection is non-empty, but does not form proper subsets). The preceding analysis argues strongly for the last of these five (5) options since it has been demonstrated above that significant overlap exists, albeit not in a properly contained manner. Figure 1.1 on page 6 is provided to describe the general relationships determined from this analysis.

The specific nature of the noted 'violations' to formal human reasoning is not unexpected given that all of these result from generally well-known issues attached to Christology and/or the Trinity, though the fact the 'violations' very generally resulted from definitions was not anticipated. The observation that the 'violations' are largely, but not entirely, separate from the 'exceptions' is indicated in the region of Figure 1) exclusively in 'Christ's Use of Conditional/Procedural Forms' above. The particular nature of the noted 'exceptions' is more varied in that there are four (4) different theological issues as analyzed in detail above, not just the two (2) theological issues that must be addressed with the 'violations', though the two (2) common issues between the 'violations' and 'exceptions' pertain to the Trinity and Divine Protection of the disciples.

Where there is commonality of the theological issues between the 'violations' and 'exceptions, only the issues related to the Trinity have fundamentally different underlying factors. Specifically, the issues with the Trinity that cause 'violations' are essentially physical impossibilities, while the issues of hierarchical authority specific to the Trinity result in 'exceptions'.

Besides the general conclusion summarized in Figure 1.1), the only three (3) unresolved, main findings from this study that cannot be mitigated via translational differences or via the assumption of literal interpretations only is that there are:

1) There are a total of thirteen (13) individual and four (4) general types of internal, theological differences affecting 'exceptions' within the words attributed to Christ. The most frequently noted, unresolved issue from the 'exceptions' results from the nature of the Trinity. The main unresolved issue *uniquely specific* (^) the 'exceptions' results from the nature of human

"belief" as such cannot be explained within the assumptions of this study and ultimately point to the critical questions surrounding human free will and Sovereign election.

2) From a strictly formally logical perspective within human reasoning, there are four (4) specific rules that cannot be applied in all cases to the words attributed to Christ. These all ultimately point back logically to the observation that multiple truth values are at times permitted in the words attributed to Christ or equivalently that the rule for 'double-negatives' is not always applicable to the words attributed to Christ. The other three (3) rules affected are: logical equivalence between statements and contrapositives, dual-inclusions implying logical equality, and transitivity of equivalence relations. Taken together, these fundamentally logical issues account for all of the noted 'exceptions'.
3) There are fifty-one (51) 'violations' of formal human reasoning noted when comparing such to the words attributed to Christ. Like the 'exceptions', the most frequently noted (82%), unresolved issue from the 'violations' results from the nature of the Trinity (^), which are part of the larger category of 'violations' resulting from physical impossibilities, though there are four (3) other specific types of unresolved 'violations' ranging from: unconfirmable statements pertaining to 'life after death'; conflict of formal rules of human reasoning; and historically verifiable conflicts pertaining to the mortality of the original Apostles of Christ. Very notably, a significant plurality of these 'violations', though not all, are found in the Gospel According to John.

(^)—The main unresolved issue applying to both 'violations' and 'exceptions', aside from the Trinity, results from two (2) promises specific to Divine Protection of the disciples. From all of this, can one say that the words attributed to Jesus Christ in the Holy Bible are 'reasonable'? This analysis indicates that an unqualified 'yes' is not possible, particularly in view of the observation that specific logical rules of formal human reasoning are not always applicable, especially the axiomatic rule for 'double negatives' which is crucial to the 'exception' related to the nature of human belief, and also given the 'violation' specific to Divine Protection. However, it can be stated from this evidence that these words attributed to Christ are very largely consistent with formal human reasoning and as such have very specific patterns for the noted 'exceptions' and 'violations'.

As a result, one can answer the primary question here with an answer of 'mainly yes', though with very particular caveats.

Where shall these findings lead? While no general claims are made outside of the context of the specific questions posed here, the general consistency and patterned nature of the noted exceptions primarily lend toward a greater impetus for even more comprehensive examination for the rest of the Holy Bible using this formal method of comparative analysis. It also leads to the general conclusion that any interpretation or analysis of Christ's Words cannot rely in all places upon formal human reasoning since in those places where there are differences all of the rules of formal human reasoning are violated at least once.

Chapter Five

The Axioms of the Logic Employed by Christ Involving Humanity: The Indispensable Christian Virtues

From analyzing all of the 759 conditions for which mankind in general or an individual is the primary entity for a desired action and/or attitude (*) in the conditional/procedural statements attributed to Christ in the Holy Bible, there are a total of fourteen (14) conditions which imply all of the conclusions in these particular type of statements attributed to Christ in the Holy Bible. As such, these 'axioms' are 'unimplied conditions' and in that sense are shown to be 'irreducible'. The conclusions associated with these conditions can either be a man's and/or God's action or attitude. These fourteen (14) conditions are thus effectively the 'axioms' upon which all of the other statements attributed to Christ in the Holy Bible are founded:

THESE FOURTEEN (14) AXIOMS

1) Sacrificial use of a believer's talents, time, and resources (including one's personal, physical well being via "persecution" and "cross-bearing" if required) to benefit other persons (believers and non-believers) for the specific purpose of growing or advancing the "kingdom of heaven" or "kingdom of God"
2) Desire for Christ's Will that is intense and most personally important for the advancement of the kingdom of God/kingdom of heaven and for personal righteousness (see Matthew 5 and 6), which is more than just persistence, though there is an aspect of persistence in this type of desire
3) Persistence in all 'axioms'
4) Faith in Christ
5) Humility in attitude and/or in action toward God and other persons

6) Applying God's Standards to assessing the behaviors and words of others and oneself
7) Repentance (or 'completely turning') from all sinful actions and thoughts with no attempt to 'hide' or 'justify' such sins
8) Alertness to (watching) spiritual matters, including temptation and to Christ's Second Coming
9) Forgiveness by a believer toward all others of their sins against or affecting that believer
10) Making disciples of Christ with sincerely interested persons by life example and/or by sharing words
11) Sincere recognition of the worth of God's Attributes and His Actions
12) Praise expressed to God and others for God and Christ's Character and Actions
13) Peacemaker with other persons
14) Patience or waiting upon God's Timing for an event to occur

(*) – KEY NOTES REGARDING ASSUMPTIONS

A) Given the lack of a formally derivable or explicitly stated deductive basis for the logical forms employed by Christ in the statements attributed to Him in the Holy Bible, no assumption is made here that these 'axioms' are absolute or formal primitives. This note is required as well since these 'axioms' are derived by inductive means through examination of the specific instances involving conditions relative to mankind. Further, to reinforce this point, 'axiom' is qualified by single quotations (via 'axioms').

B) It is recognized in all of these axioms that humans cannot *solely* initiate or continue in meeting *any* of the attitudes or actions that constitute these conditions. Indeed, John 6: 44 and 65 both indicate that this assumption is explicitly part of the statements attributed to Christ. Even more explicitly for the purposes of the analysis in Chapter 5) of this study, God the Father initiates such 'desire' upon which the human 'will' must in turn respond as per assumption in Chapter 5) and indeed, as per John 15: 17 ("ye have no chosen me, but I have chosen you . . .") and John 6: 65 (". . . no man cometh unto me, except it were given unto him of my Father"). In so doing, this application in Chapter 5 agrees with the theological assumptions from the first four (4) chapters, though the assumption labeled 1.d) is significantly extended in the preceding statement as in Chapter 5 this assumption applies to all of the 'axioms'.

GENERAL NOTES

It has well noted that the use of 'transitive' reasoning (that is, "If A, then B and If B, then C implies that If A, then C) is only employed once explicitly in the conditional/procedural statements attributed to Christ in the Holy Bible. So, this application must be so cautioned as to not assume that Christ explicitly attempted to identify such 'axioms' as a general practice. However, this application does not differ with any of the conditional/procedural statements attributed to Christ nor does it violate any of the formal rules of human reasoning as the only logical assumptions made in this application are that all of the conditional/procedural statements are true within the logical framework employed by Christ (and as shown in this analysis with only a few exceptions to formal human reasoning) and that the transitive property of conditional statements can hold. No uses of other logical forms (equivalences, contrapositives, negations, etc.) are required for this application detailed in Chapter 5. Moreover, none of the statements associated with the 'exceptions' and/or the 'violations' from the analysis in Chapters 1-4 are required to identify these 'axioms' since the conditions and conclusions associated with the 'exceptions' and/or 'violations' are not considered on the basis of the logical issues here and also practically since it so happens that the conditions and conclusions associated with the 'exceptions' and/or 'violations' would not change the 'unimplied conditions' or 'axioms' so noted in Chapter 5.

A few of the 'axioms' are very direct in terms of linkage between the pertinent conditions and conclusions. For instance, Christ is indicated in the Gospel of John (chapter 14, verse 15) as saying, ". . . if a man love me, he will keep my words" (call this example # 1). There are also though even more statements attributed to Christ that require multiple implications (via the 'transitive property of differencing'), which as noted previously are not explicit in the Holy Bible, but for which there is neither any significant evidence to suggest that cannot be employed. An instance of such multiple implications tracing back to the 'axioms' can be seen by connecting example # 1 from just above with another statement attributed to Christ in the Gospel of John (chapter 15, verse 10) as follows: "If ye keep my commands, ye shall abide in my love . . ." (call this example # 2). Thus, by linking example # 1 and example # 2, "love" (of Christ) implies not only keeping of (Christ's) words, but also a resultant 'abid(ing)' for that person who loves Christ.

It should be noted that care has been taken to insure that contexts and original Greek wording are all aligned/matching where multiple implications are considered in determining these axioms.

Some possible concerns with this list of 'axioms' need to be addressed. Specifically, four (4) pertinent questions that may be asked of this list are as follows:

1) Why is prayer not part of this list of 'axiomatic' conditions? The basic answer is that effective prayers or effective praying in the Holy Bible are ultimately conditioned by one of the other traits in the list of axioms above, except for 'alertness to temptation and Christ's Second Coming' (#). Notably, this is a valuable insight in and of itself given that so much is said in the Holy Bible regarding the importance and spiritual power of prayer. As such, effective praying is not separated from these 'axiomatic' conditions and should be seen not as a virtue or trait to be pursued for its own sake, but is rather a definitely implied trait of those who meet the 'axioms' listed above. More directly, such analysis of Christ's Teaching on effective praying/prayers reveals that the specific conditions for successful praying are the same as those upon which all of the other conclusions/promises from Christ's Teaching also rest. As an aside, only 'alert to temptation and Christ's Second Coming', 'applying God's Standard', 'sincere recognition of God', and 'witnessing' are not explicitly stated in the Holy Bible as conditions for effective praying, but are considered 'axioms' since there are no conditions to imply these two 'axioms' and since all of these 'axioms' are directly stated as vital parts of Christ's Command to 'abide in His Words' in order to receive what one is asking God to do within His Will, such as noted in John 15: 7. Thus, according to the logical forms used by Christ, effective prayer results from a life committed to the axioms above; and more practically, this analysis demonstrates that not only are there specific conditions to effective praying, but that praying effectively is essentially the 'outgrowth' or 'extension' of a believer's life who is dedicated to these 'axioms'.

(#)—'Alertness to temptation and Christ's Second Coming' is a separate axiom from all other 'axioms'. This includes conditions in addition to 'pray(er)' since all of the conditions, associated with conclusions in the conditional statements that are attributed to Christ in the Holy Bible that explicitly command 'alertness' or 'watch(ing)', are specific only to Christ's Second Coming or to human temptation, which is not within human control or even fully within human knowledge. Also, such 'alertness' or 'watching' is explicitly distinguished in five (5) total conditional statements attributed to Christ in the Holy Bible as either a condition or as a command/conclusion that is not implied by or associated with any of the other 'axioms' and since 'watch' is an explicit condition in six (6) statements attributed to Christ in the Holy Bible, with two (2) of these six (6) instances being associated also with a condition or 'prayer'.

2) Why is personal righteousness or obedience (not including those statements where the context involves the "righteousness of Christ" as is always conditioned to those who have 'faith' in Christ) not part of this list of axiomatic conditions? The answer to this very pertinent question is that 'obedience', 'righteousness (not specifically 'of Christ'), 'keeping commandments', or 'following Christ', etc. is what may be called an *encompassing* term for all of the axioms, except for 'faith', where Christ stated repeatedly in Matthew 5-7, amongst others, that such 'obedience' is not equivalent to a literal adherence to the 'letter' of the Mosaic law nor to the 'traditions' of religious leaders, but rather is indicated as being the 'fulfillment' of the Mosaic law in terms of focusing on the conditions of one's 'heart' by an adherence to the 'spirit' of the law.

{Note: The 'axiom' of 'desire for God and Christ above all' by mankind is assumed to be present in all positive conditions for the promises attributed to Christ since without such 'desire' none of the other conditions can be fully satisfied. Ultimately, this view points back to the assumptions of free volitional acts by mankind, but yet also spiritual 'desire for God and Christ' being initiated by mankind. The main question then is whether total or 'pre-eminent' desire for God and Christ that is above all other desires by a human who would realize the conclusions promised by Christ. This question of total, 'pre-eminent' desire for God and Christ is addressed as follows: In Matthew 6: 24 and 6: 33 as well as Mark 9: 34-35, total commitment to none other than Christ and 'deny(ing)' of selfish, personal interests is stated as being a necessity and is reinforced strongly by such statements as "No man can serve two masters . . .". While the analysis in Chapters 3) and 4) demonstrated that a strictly 'binary' interpretation of Christ's words cannot be supported because of noted 'exceptions', only one (1) of the 'exceptions' indicate other than a 'binary' view of Christ's words, that is of those that affect the question of total commitment to Christ. Namely, even though 'Exception 1.a)' for "good(ness)" cannot be fully resolved, which results in even disciples of Christ not being absolutely in all cases and at all times "good", 'desire' being totally committed to Christ is not reduced by "good" inherently since "good(ness)" implies both actions and attitudes, where 'desire for God and Christ being pre-eminent' is fundamentally an attitude. Hence, this question is reduced to whether one can be "good" in one's attitudes or 'heart' and yet not completely "good" in action, which has been established as possible from the analysis in Chapter 3) and 4).}

A legitimate question then would be: why are there not only two (3) 'axioms: namely, 'faith', 'desire', and 'obedience'. The direct answer is that 'obedience', could be used as essentially one (1) of only three (3) axioms, but the other 'axioms' (not including 'faith') provide a fuller, more specific explanation of the fundamental 'axioms' of the logic employed by Christ since each of these other axioms all imply conclusions/promises involving 'obedience' of some form (such as those indicated above).

3) Why is 'making disciples' an axiom, but 'discipleship' is not an axiom? "Discipleship' of oneself is not the same as a believer 'making disciples' of others. 'Making disciples' is specifically presented in the statements attributed to Christ as part of the same continuum of action as is initiated by 'witnessing' on the part of those who already are following Christ, and both 'witnessing' and 'making disciples' share the stated condition that neither actions should be used with persons who demonstrate repeated, complete unwillingness to respond to such efforts. As such, these two (2) conditions should not be viewed entirely separate or as distinct concerns, as often seems to be case. Notably though, in a strong sense when considering procedural timing, 'making disciples') always is preceded by 'witnessing' (though not necessarily by the same individual importantly). However, there are no logical forms in the statements attributed to Christ in which 'witnessing' implies 'making disciples' (as defined here) or vice-versa, where 'making disciples' is used to mean the specific action of aiding others to follow Christ in their own life of 'faith' and obedience to Christ, not the act or attitude of following Christ oneself. Also, while inherently tied to one another in time particularly, 'witnessing' about Christ to someone (even by Christ Himself) is always used to describe interactions with non-believers, 'making disciples' is generally tied to interactions involving only believers (both the 'discipler' and the one being 'discipled').By way of example, when Christ sent the seventy (70) disciples to be 'witnesses' of Christ in Luke 10, He did not address the aspect of 'making disciples'. Critically though, in one (1) place Christ specifically commanded the twelve (12) apostles to "make disciples" (see Matthew 28), though in all of the cases that 'making disciples' is used as defined here (i.e., "making disciples") there is either explicit indication that 'witnessing' is first required in order of time (such as in Matthew 28) or there is an implicit assumption that 'witnessing' has already occurred in view of the context of the statements for which "making disciples" is a part of a conditional statement. Finally, there is clear evidence in Matthew 3 that Christ's only condition to "witnessing" or to "becoming fishers of men" was for one to "come after" Christ or to "follow" Him, which is conditioned upon other 'axioms', including especially 'sacrificial use'. Thus, 'making disciples' implies 'witnessing' from a conditional, logical perspective, though 'witnessing' precedes 'making disciples' in time.

4) Why is 'fast' not an axiom since it has such vital importance to realizing spiritual benefits? The act of 'fasting' is specifically mentioned as a prerequisite condition towards realizing very specific results/promises, such as

in Matthew 17: 12 for certain types of spiritual activity, such as casting out certain demons. More generally though, 'fasting' is seen as specific case of a more generalized condition in the statements attributed to Christ involving personal sacrifice on the part of one who was considering or was already following Christ so that 'sacrificial use of one's time, talent, and resources' implies 'fasting'.

5) Why is 'persistence' a separate axiom from 'faith'? While it is true that there are many instances in the Holy Bible in which persistence through waiting or through trials is a key element of an individual's 'faith' in God and Christ, the axiom of 'persistence' has a distinct sense of timing for procedural, logical forms. For instance, in Luke 18, 'persistence' is a positive condition for effective prayer, but there is no explicit specific sense of 'faith' or even more generally 'faith' where such 'persistence' is required. There are, however, a number of instances of conditions involving phrases such as: 'abide in me (Christ)'; 'my (Christ's) words abide in you (a disciple)'; or 'abiding in my (Christ's) words', and all of these conditions are implied by the more general condition of 'persistence'. Since 'faith' (in Christ)' is an 'axiom' and as is part of 'abiding in Christ', the 'axioms' of 'persistence' and 'faith' have a definite overlap in this regard, but neither is fully implicative of the other and hence are not the same condition.

This validly though raises the substantive question of whether true 'faith' may ever become non-persistent. There are nine (9) conditional statements attributed to Christ that must be considered relative to this topic—specifically, John 15: 6; Luke 9: 62; Matthew 24: 13; John 8: 31; Revelation 2: 11; Revelation 2:26; Revelation 3: 12; and Luke 8: 13. First, John 15: 6, if taken literally, states that lack of a man abiding in (Christ) will result in that man being "cast forth as a branch . . . and burned", which, if taken literally, is a reference to a believer who does not persevere to eternal life. Second, Luke 9: 62 addresses this question directly since it involves use of an 'inverse' logical form ("no man having put his hand to plough.. is worthy of the kingdom of God"), which is the 2nd most strongly matching statement of these nine (9) to the manner in which the question is posed here (see the final one in Luke 8: 13 for the strongest example). Third, Matthew 24: 13 explicitly states from a literal interpretation, that 'being saved' is contingent upon "endur(ing) to the end". It is recognized though that Matthew 24's context is specific to "signs of the end" of time, so that this particular statement may be further conditioned to the times and persons included there (depending on various eschatological viewpoints). Even so, fourthly, the context of John 8: 31 (which states that ". . . If ye continue in my (Christ's) word, then are ye my disciples indeed . . .") is specific only to Jewish believers in Christ, which is considered in this manner applicable to all believers in Christ. However, this is the 'inverse' of the manner in which the question is posed here since the condition does *not* say "if ye continue not in my word, then are ye not my disciples". This is important since it has been shown that the 'inverse' of a statement is always necessarily true, so John 8: 31 is not deemed directly applicable to this question. The fifth instance pertains to Revelation 2: 11 in which "overcometh" is connected to "not be(ing) hurt of the second death". Since the term, "second death", is not defined in the words attributed to Christ, it cannot be assumed via literal interpretations only within the scope of this book that the "second death" is effectively the opposite of "eternal life", though notably in Revelation 20: 14, a definition to this effect is provided. The sixth instance occurs in Revelation 3: 5 in which "overcometh" is linked to "not blot out (one's) name out of the book of life . . .". Even though the use of verb tense from "eth" verbs is not assumed here, the context is assumed from the use of "overcometh" to imply 'persistence of faith'. However, this passage does not support from a literal perspective the view that 'true faith may not always persist' for the same reasons as the fourth instance here just above. The seventh and eighth instances of this type are found in Revelation 3: 12 and Revelation 3: 21, respectively, and are essentially similar to the sixth instance, though the conclusion in the seventh example here states, "a pillar in the temple of my God, and he shall go out no more" and "will I (Christ) grant to sit with me in my throne", respectively, which though they have a similar connotation of indicating "eternal life" are not in the form of the inverse as needed to be applicable explicitly to the question as posed. Finally using an illustration, in Luke 8: 13 those who receive the 'seed' of the 'word' and 'believe', but are on 'rock(y) ground' and 'fall away', is the most specific and strongest example of those using a literal interpretation who 'receive' and 'believe' (in the word, which literally can be taken as Christ), yet who 'fall away' and such do not persist, to the question of non-persistent 'faith'.

Before concluding the analysis of this question, it is noted that are four (4) statements attributed to Christ in Revelation 2: 7b, 10, 17, and 26 (see # 1573, # 1574, # 1578, and # 1586 in Appendix A) is included by some in the list of verses indicating that 'true faith may not be persistent'. However, since the conclusions of these particular statements literally state "eat of the tree of life", "crown of life", "eat of hidden manna, and will give him a white stone, and in the stone a new name . . .", and "power over the nations" respectively, not "eternal life" or a very similar phrase, the assumption of this study regarding literal interpretations only do not permit the usage of these four (4) instances to address the question of 'true faith' persisting or not.

Having noted that there is evidence to support 'true faith not being persistent' if Luke 9: 62 and John 8: 31 are taken literally, it is quickly recognized that this only addresses the 'what' of this question, but in manner address the 'how' or the means by which such 'persistence' occurs. Ultimately, this points to 'perseverance of the saints under Sovereign election' and 'free will'. Using the two (2) 'additional assumptions' in Chapter 5, only 'perseverance of the saints under Sovereign election' requires further examination since 'Sovereign election' is a much more specific assumption than that God's grace, power, and action are required to initiate and sustain human spiritual activity.

6) Do any of the conclusions/promises attributed to Christ in the Holy Bible require combinations of these 'axioms' as conditions in order to realize a particular conclusion/promise? While the majority of the conditional/procedural forms attributed to Christ in the Holy Bible require only one (1) of these axioms to be realized or satisfied (once 'faith' and 'desire' can be assumed to be present in the life of a believer in Christ), there are more than a few conclusions that require more than one (1) of these axioms to be realized. For example, in Matthew 5, when Christ told his disciples to fast, but to do so in a private and 'humble' manner in order to realize the "reward" that God would give to them for the sacrifice of fasting, two (2) of these axioms (i.e., 'fasting/sacrificing' and also 'humility') are required as conditions for the promise to be fulfilled. Additionally, 'persistence' combined with 'forgiveness' is a must: see Matthew 18: 21-35 for parable of the talents owed by servants to their lord. Very importantly as well from 1) above, effective praying is strongly a combination of all of these axioms.

7) Why are actions and attitudes of humans both considered as axioms? One of the clearest implications of the statements attributed to Christ is that actions (both "good" and "evil") are closely associated with corresponding attitudes (for instance, see Matthew 15). Notably, all but four (4) of the axioms here are focused on 'attitudes' of a believer. The four (4) axioms not adhering strongly to a general focus on 'attitudes' are for the axioms of 'faith', 'sacrificial use of time, talents, resources . . .', 'witnessing about Christ', and 'making disciples', based on the contexts associated with these four (4) conditions in the statements attributed to Christ. The corresponding 'attitude' for these more action-oriented axioms is what shall be called here 'acts of the will' in which one's 'will' is seen as essentially the nexus of all human beings' tripartite nature, which is composed of body, soul, and spirit. (Aside: It is beyond the scope of this analysis to examine the precise relationship between the three parts of mankind, especially in regard to attempting to parse the relationship between the 'heart', the 'mind', one's 'strength' and the body, the soul, and the spirit.)

Additionally, a related, very valid, question is whether any of the more attitudinal-related 'axioms' imply or have strong connection to other 'axioms' in the words attributed to Christ in the Holy Bible. The answer to the former is a strong 'no' regarding one 'axiom' implying another 'axiom' since, by definition, any such 'so-called axiom' implied by another 'real axiom' would be subsumed as a result of the 'real axiom' and hence would not actually be an 'axiom', but regarding the latter question as to whether there are strong connections between the 'axioms', the answer is a definite 'yes'.

First in this regard, the 'axiom' of 'desire for Christ's and God's Will' is strongly connected to all of the 'axioms' here, for without such a 'desire' on the part of a believer, no condition required for a believer to obtain God's Promises in the words attributed to Christ in the Holy Bible can be realized {see the Note for 2) above}. Also, see Mark 12: 41-44 for the words attributed to Christ regarding the poor widow giving two (2) mites, who is commended as giving more than all others at a time of offering, not sacrifices, for her giving being truly sacrificial. This is particularly so for 'sacrificial giving of one's talent, time, and resources', since no such self-giving behavior can be spiritually productive without such a 'desire', which Christ exemplified in all of His Actions in the Holy Bible, and which Christ told his disciples is required to "follow after" Him in Luke 9: 23 and Matthew 16: 24. Notably though, nowhere else does 'desire' directly imply 'sacrificial use' since 'desire' must be expressed in an outward, tangible manner to be realized, such as commanded in Matthew 6: 33 where "seek(ing) . . . first the kingdom of God and His Righteousness.. (results) in all these things be added (to a believer) . . .".

Further, such self-giving 'sacrifice' is the most frequently noted of 'axioms' in the words attributed to Christ in the Holy Bible and implies many of the other positive conditions for a believer, such as 'agape love' of God and others (see Matthew 6: 21 and Matthew 7: 12) as well as 'good fruit' and 'obedience (of Christ and God)' (see Matthew 7: 17-18). Additionally, such 'desire' is clearly indicated in the words attributed to Christ in the Holy Bible as being necessarily the most important 'desire' in life, to the point of Christ saying in Matthew 6: 24 and Matthew 12: 30 effectively that there is no 'middle ground' in the life of a disciple of Christ in regard to His allegiance to Christ. {Technical Note: Such 'desire' is not the same as 'good intentions only', for without any demonstration of that 'desire', such 'intentions' are strictly just that in the life of a disciple of Christ.}

Such commitment does not in any manner though imply that 'desire' will not deepen and grow with increasing experience, understanding, and knowledge as a maturing believer in Christ to which Christ said in Luke 12: 48, ". . . to whom much is given, much shall be required." It may well be asked here why such 'desire for Christ's Will' is strongly connected to the 'axiom' of 'faith' in Christ' since the analysis in 2) above recognizes that such 'faith' is fundamental to the Christian life. In order to address this, it must first be remembered that as just noted above, such 'desire' is only required to match the present level of Christian experience, understanding, and knowledge as he/she matures in Christ. Given this, such 'desire' for a 'new-born' believer in Christ does not have the same depth or breadth of 'desire' as that to be required of believer who is no longer a 'babe in Christ'. Based on connecting John 8: 12 (in which "hav(ing) the light of life" is conditioned upon "following" Christ) to John 12: 26 (in which "follow(ing)" Christ is conditioned upon "a man serv(ing)" Christ), there is an explicit connection using the conditional statements attributed to Christ between "serv(ing)" Christ and "hav(ing) the light of life", which is taken in the contexts of John 8: 12 and John 12: 26 to mean 'eternal life' for 'the light of life' since Christ said that "I (Christ) am the light of the world" and hence also the "light of life".

Such 'serv(ice)' of Christ must be preceded by a 'desire' to do so, which again God initiates, but which the true disciple must cultivate and express in actual fact. Also, the story of the "rich, young ruler" in Luke 18: 18-25 (in which Christ conditioned eternal life upon a true willingness for this young man to place "following Christ" above his own desires, which in this particular case involved 'riches'), reveals that in order to "have eternal life" one must have at least a 'willingness' to please Christ in a self-giving, sacrificial manner to the extent that the spiritual experience, understanding, and knowledge which they possess at the time of expressing 'faith' in Christ in order to have eternal life. The above analysis also addresses a key aspect of the long-standing question of the salvific relationship between "'faith'" and 'works' by indicating that it is not the actual 'works' themselves which are salvific, but rather the 'willing desire' as described above to do the 'works' that demonstrates true 'faith', for any 'works' attempted for their own sake are merely 'filthy rags' in the sight of Christ. So, 'faith' in Christ which is devoid of any 'willing desire' to please and serve Christ is not true 'faith'. Moreover, in Matthew 5: 20, works for works' sake is not adequate for one to ". . . enter into the kingdom of heaven".

Having addressed the relationship of 'desire to do Christ's Will' to the other 'axioms', there is another key relationship between the attitudinally related 'axiom' of 'humility' and the other 'axioms' here. This relationship is of a similar quality to that of a 'desire to do Christ's Will' in that there is an inherently unselfish serving aspect to all true attitudes and actions associated with 'humility'. Such 'humility' is required for all Christian discipleship when Christ said in Matthew 18: 4 being "called the greatest in the kingdom of God' is conditioned upon being "humble" and again in Matthew 20: 27 where 'activated humility' or "serv(ice)" results in a similar conclusion of being "chief among" (the disciples). True 'humility' is though different from a 'desire to do Christ's Will' in that such 'humility' originates from personal knowledge of who God and Christ are in their Divine Nature and in their Divine Acts in relation to a knowledge of the limitations and 'sinfulness' of man, while 'desire to do Christ's Will' is an act of the 'will' of a believer, which as stated before is the 'nexus' of the body, soul, and spirit of a person. This is not meant to imply that 'desire' and 'sacrifice' have no overlap since the definitions of these terms here require such connection given that the 'will' of a human being will always intersect personal knowledge. However, the 'will' of a human will not always be fully consistent with personal knowledge since otherwise humans would never 'will' to act in ways that are inconsistent with personal knowledge, which is definitely not the case, even with believers as is demonstrated in multiple places even with the disciples, such as with Peter's denial of Christ.

8) Why is 'love of Christ', 'love of God', or 'love of others' not listed as an axiom? It is definitely true that the phrases: 'love of Christ'; and 'love of others', are connected with many statements attributed to Christ in the Holy Bible, especially in John 14, and the Holy Bible has much to say about the vital importance of true 'love for Christ' and 'love of others' (such as in I Corinthians 13 and I John 2). However, having examined all of the instances in which 'love of Christ' or 'love of God' is a condition to a conclusion/promise, a part of, or all of a definite conclusion/promise, the only consistent pattern that emerges is that of 'sacrificial use of one's time, talents, and resources' being the greater antecedent of the type of "agape love" (which is a sacrificial act of one's will in a Godly manner by "deny(ing)" oneself—see Luke 9: 23.). In John 14 (see verses 21 and 15), there is a key instance that is indicative of a much broader principle, in which Christ is noted as saying that 'agape' love for Christ both implies and is implied by one's personal 'keeping of (Christ's) commandments'. As noted in the analysis from the preceding chapters here, this set of verses constitutes a unique construction of logical forms in the statements attributed to Christ in that a statement and its converse (using the terminology of formal rules of human reasoning) are attributed to Christ. Moreover, by connecting the condition/conclusion pairs of Matthew 6: 21 ("where one's treasure is, there shall one's heart be also") with such statements as found in Matthew 7: 17-18, it is seen that not only is 'sacrificial use of one's resources' implicative of the condition of one's "heart", but also that such 'sacrificial use' will by further implication also result in "good fruit" (including "keeping of (Christ's) commandments"). More directly, personal sacrifice is evidenced as the primary condition for "being a disciple" (not "making disciples") such as is found in Matthew 16:24. This insight also addresses a very fair question regarding why 'purity of heart' is not its own separate axiom, even though 'purity of heart' is a key attitude commended by Christ, especially in Matthew 5: 4. It is critically important though to note that 'sacri-

ficial use of one's time, talent, and resources' is implicative of what many call the 'Golden Rule' (see Matthew 7:12) and as such is further, strong evidence of the vital importance attached to the axiom for such 'sacrificial use' in relation to "agape love" of others, not just God and Christ. Finally, such 'agape love' is required to extend even to those who are 'enemies' toward a believer in Christ so that the conditions for "love your enemies" are satisfied as well by the 'axiom' of 'sacrifice'.

9) Why is speaking in a straightforward, yet encouraging manner not an axiom? While speech is very important, the condition of one's heart is even more important (see Matthew 15: 18-19), and indeed this passage reveals that the condition of one's 'heart' is directly implicative of the manner of one's speech. Speaking in a straightforward, yet encouraging manner is also seen as part of the broader call for believers to be 'transparent' in their motives, attitudes, and actions. Such 'transparency' is seen as ultimately redounding from one whose heart is purely focused on knowing and pleasing Christ above all and as such whose primary aim is to hear what this author considers to be ultimate blessing of our Lord, "Well done thou good and 'faith(ful)' servant, thou hast been 'faith(ful)' over a few things, I will make you ruler over many things, enter thou into the joy of the Lord. (Matthew 25: 21, 23)"

10) Are there any additional 'conditions' associated with the axiom of 'repentance' besides the other axioms listed here? Yes, there is one (1) other condition associated in the statements attributed to Christ in the Holy Bible with 'repentance' (or more specifically, receiving forgiveness, which is a direct result of true repentance). This condition is noted in Mark 3:29 in which such, true repentance and true forgiveness are conditioned upon not having "blasphemed the Holy Ghost" or "speaking against the Holy Ghost". Since this phrase, "blasphemy of the Holy Ghost" occurs in no other contexts than those essentially the same as Mark 3:29, no further analysis within the literal confines established here is able to be pursued without appealing to interpretive measures beyond the two (2) theological assumptions as well as the one (1) requiring 'literal meanings only' employed in this book.

11) Does not the entire premise of the establishment of such 'axioms' and more generally of conclusions/promises being realized from the satisfaction of specific conditions promote what is generally referred to as "transactional theology" in which to speak plainly, "if man does X, then God promises to do Y"? This is a fair question from a technical perspective theologically. However, it is simply, but respectfully, admitted by the author of this analysis that a strictly literal rendering of the statements attributed to Christ in the Holy Bible can reveal no other viewpoint than that a preponderance of the actions from or by God or another member of the Trinity are conditional (approximately 50% of which involve an action or attitude on the part of mankind,

though objectively this percentage is significantly lower if one considers only those forms in which the words "if", "since", or "when" are explicitly stated).

This strict use of literal meanings is not though merely a "safe" position from dealing with the serious questions raised by other non-literal renderings of these statements, but is rather a result of the intended analysis here (namely, to compare and contrast the logical forms of Christ to the formal rules of human reasoning in which non-literal meanings have no technically acceptable value). Lastly in this regard, it should be noted that there is no basis from this analysis for concluding that man's not meeting the specified conditions will always result in God's Promise not being realized (as is shown in the discussion on 'inverses' in Chapter 3) or for reasoning that if a member of the Trinity does or does not act on a specific promise that specific persons necessarily will or will not meet the condition required (as is also shown in the discussion on 'converses' and 'contrapositives' in Chapter 3).

12) Why are there no axioms involving multiple persons all of which are acting to meet a specific condition or conditions? This is a very valid question, and one which has a response that was not anticipated prior to this research by the author. There are three (3) general cases to address initially as there are three (3) 'axioms' which involve in the most general sense at least one (1) other human being beside a particular believer. First, the 'axiom' for 'forgiveness' of others' sins' involves multiple persons, but only from the perspectives external to a believer against whom a sin has been committed. It is also well noted in this specific case that Christ did not say that the 'offending' person (who may or may not be a believer) must 'accept' the 'forgiveness' of the believer who offers such forgiveness in order for the 'forgiveness' to be acceptable by God, though if an 'offending' believer "repents", the other believer is required to repent according to Luke 17: 3. Thus, in this first specific instance of 'forgiveness of others sins', multiple persons are involved, but these persons do not have to act physically or spiritually for the positive conclusions with such 'forgiveness' to be realized.

Second, a similar instance applies to 'witnessing' in that 'successful witnessing' does not require 'acceptance' by the person receiving the 'witness about Christ'. Third, while 'applying God's Standards to assessing other's behaviors or words' does technically involve multiple persons, the 'application of God's Standard in assessing others' behaviors or words' does not directly require any interaction with or participation by any other person aside from a believer who is assessing another's behavior or words in this manner. There may well be interaction with and/or participation by or impact to another person if action is taken based on this assessment (such as, with believers according to Christ's principles of 'conflict resolution' within the church as given in Matthew 18: 15-17 or, for another example, in the case of either a believer or possibly a non-believer when there is an "offense of one of these little ones" in Luke 17: 1-2), but such assessments do not always involve or affect others directly.

Now, having addressed the more direct, general cases for the 'axioms' involving multiple persons, the following will consider any conditions (including the other eight (8) axioms and any other conditions implied by these 'axioms') in which the primary action is conditioned upon 'joint' human activity and/or attitudes:

i) Matthew 18: 19-20: "If two of you (disciples) shall agree on earth as touching any thing that they shall ask, it shall be done for them of my Father which is in heaven". This is not considered an additional set of conditions and hence not a 'new' axiom. This is so given that the conditions of effective praying are already addressed by the existing axioms as seen in item 1) above and since to recognize such "agreement" by two (2) or more disciples of Christ as a separate set of conditions from individual prayer would imply that there are types of spiritual activity that can only be achieved by "combined" prayer in "agreement", which is a view that cannot be supported by another statements attributed to Christ in the Holy Bible (nor more generally from the rest of the Holy Bible). All, but three (3), of the conditional statements attributed to Christ in the Holy Bible associated with a prayer or praying, which do not involve a specific person given by name, such as Christ or the importune/persistent widow, rather involve groups of persons (generally by the use of the plural form of the 2nd person pronouns, "ye" or "you").

   Indeed, in the rest of the New Testament, all instances in which prayer, not specific to a particular, named person, are conditioned upon plural pronouns, which involve multiple persons/groups of persons. It should, however, be noted from analysis of all uses of the singular, plural, and indefinite pronouns (ye, you, thou, thee, whosoever, and whomsoever) that the three (3) passages in the Holy Bible not fitting the general pattern above (and, in fact, in the New Testament) having the following in common: a) no individual persons are listed by name (which would not include any passages specific to Christ praying by himself as He often did); b) the subject of the condition (which in both cases is 'faith' or belief) involves particular individuals (as opposed to a group of persons); and c) the conclusion of the conditional statement is associated with prayer (see Mark 11: 23, Luke 17: 6, and also Matthew 6: 6. Only Luke 17: 6 uses a noun other than "thou", which is very generally the singular form for the

2nd person pronoun. Luke 17:6 uses the plural form "ye" in the King James Version, but this statement also has the indefinite form present that is usually translated in the New Testament as "whosoever").

Interestingly, Mark 11: 23 and Luke 17: 6 both involve a very similar context of Christ teaching the disciples about praying to ask that unproductive or spiritually hindering objects, presumably based on the context of the passages that are external to the person(s) praying, be removed. Matthew 6: 6 contains Christ's Command for each individual disciple (using the pronoun, "thou") to pray 'in private' (or "in(to) thy closet"). However, ultimately, such analysis reveals that individual prayer is commanded and given by 'successfully answered' examples, including especially by Christ, and that 'group prayer' is also commanded and given by 'successfully answered' examples in the New Testament, with only "agreement" in prayer as being additionally required when 'group prayer' is employed. Finally, Matthew 18: 19-20 is the only instance in this study involving groups or multiple persons relative to prayer where there is an assumption of individuals praying as a group, as opposed to a group of individuals who may each pray separately.

ii) Corporate worship is not addressed in the statements attributed to Christ in the Holy Bible, though such is very clearly practiced after the 'birth of the church' in Acts 2. There are groups of people who worship God and Christ in the Holy Bible, but none of these are in association with the statements attributed to Christ in the Holy Bible.

13) Why is individual 'worship' not listed as an 'axiom'? There is only one (1) type of acceptable worship in the words attributed to Christ in the Holy Bible. This is found in John 4: 23-24 in which Christ said, ". . . they that worship (God), must worship Him in spirit and in truth." Thus, 'worship' alone is not an axiom since it is conditioned upon recognizing God's Worthiness for Who He is and for His Acts in a manner that has the attitude of being 'in the spirit' and acting upon belief in the truth of God's Word. Notably, this term, 'spirit' (other than the Holy Spirit) is not part of any conditions of other than that noted in John 4: 23-24, as noted above, in the conditional/procedural statements attributed to Christ.

Additionally, the phrases ('doing the truth', being 'of the truth', or 'knowing Christ' personally as 'the truth') are noted as conditionals in the following three (3) passages: 1) John 3: 21 (those who do the truth will come to the light (the Christ in that context); 2) John 8: 32 (in which Christ conditioned 'the truth making free' upon 'knowing the truth'); and 3) John 18: 37 {being 'of the truth' is the condition for 'hear(ing) the voice (of Christ)}. {Note: John 14: 6 is a conditional/statement since it provides a definition in which Jesus simply declares that He is 'the truth', which by the rules of strict human reasoning literally implies that whatever is 'of the truth' is 'of Jesus' and vice-versa..} However, all axioms in this context apply only to those who 'know Christ' as Savior and Lord so that John 18: 37 is not an issue here. This leaves John 3: 21, but there 'doing the truth' brings one back to 'personal righteousness', which is addressed in 2) above. Finally, proper 'worship' of God as an act (such as in church) assumes that the attitude of 'in the spirit' and 'in truth' are met. So, a 'worship' in one's lifestyle actions and one's attitude precedes the actual act of worship in church. (Note: 'Worship' is different from 'sincere recognition' as 'worship' pertains to attributing worth to God's Character and/or Actions while 'sincere recognition' is specifically akin to the attitude of thankfulness and acts of thanksgiving for God's Character, or for what God has done, is doing, or will do for you and for others {see 14) below}.

14) Why is 'thankfulness' not an axiom? Thankfulness and acts of thanksgiving are forms of 'praise', and neither the word, 'thanks' or any of its main synonyms are present in the words attributed to Christ. So, the more generally question is whether 'sincere recognition' is an axiom. The word, 'sincere recognition' is not in the words of the conditional/procedural statements attributed to Christ in the Holy Bible. The only synonym for 'sincere recognition' in the conditional/procedural statements attributed to Christ in the Holy Bible is 'honour' (see John 5: 23) in which Christ conditioned 'honor' of the Son (Christ) upon 'honour' of the Father. Also, Christ did direct the disciples "when they pray(ed)" in what is commonly called 'The Model Prayer' in Matthew 6 to reverence God the Father as being 'hallowed", but this term of reverence is associated with worship as defined in 13) above. Moreover, this instance is not a conditional statement, but rather is to be a general part of all Christian prayers, which has been addressed in 1) above. There is, however, one (1) instance during Christ's 'Triumphal entry' into Jerusalem (see Luke 19: 40) in which Christ made 'praising' Him a key conditional to being a follower of Him at certain times, albeit only once.

15) Why is 'patience' or 'waiting' an axiom? The concept of 'patience' or 'wait', and its other forms as verbs, adjectives, and nouns is found across the major English translations of the Holy Bible in only one (1) place (see Luke 24: 49) from the words of the conditional/procedural statements attributed to Christ. Specifically, Christ told the disciples to "tarry . . . until ye endued with power from on high." (NOTE: The word 'patience', is employed in two (2) passages attributed to Christ using the King James Version translation—see Luke 21: 19, but multiple, English translations of the Holy Bible employ essentially the concept of 'persistence' in this passage, so Luke 21: 19 is not considered definitely to be an example with a condition of 'patience' or 'waiting'. None of the synonyms for 'patience' are found in the conditional/procedural

statements attributed to Christ in the Holy Bible.) The 'axiom' of 'patience' is considered to be different from 'alertness to spiritual matters' since 'patience' inherently implies a sense of procedural (if-then) timing from the perspective of mankind, while 'alertness' is a continual condition of the true disciple of Christ.

16) Why is 'study' of God's Word not an axiom? The word, 'study', and its other forms as adjectives or verbs are not in the words of the conditional/procedural statements attributed to Christ in the Holy Bible. Also, none of the synonyms for 'study' are found in the conditional/procedural statements attributed to Christ in the Holy Bible. There are also no conditions in the conditional/procedural statements attributed to Christ in the Holy Bible that address the concepts associated with 'study'. There is though a repeated theme of conditions within the conditional statements attributed to Christ directing persons to 'hear the word' (of God and of Christ) so that other conclusions may follow.

Thus, what does the term, 'hear the word' mean? In the context of the numerous passages in which such occurs, this concept is understood to mean definitely more than 'auditory perception' or even 'mental comprehension' of certain words. It is tempting to think of 'hear' in this sense only to mean 'understand in a spiritual sense' such as is seen in John 8: 47 and in Mark 4: 12, but the parable of the sower makes it clear that this is not always the case (see Matthew 13: 1-9 and 18-23) since 'understand(ing)' and 'hear(ing)' are in that instance separate conditions so that this type of 'hearing' is either 'auditory perception' or 'mental comprehension'. Notably, 'hear(ing) of the word' is also specifically distinguished from 'doing' or 'obeying' the word of God and Christ (see Luke 8: 21 for an example) since both 'hear(ing)' and 'do(ing)' are conditions to being one of Christ's 'brethren'. Moreover, it must be noted, from Mark 4: 11 (just preceding the reference above where 'hear' means 'understand spiritually'), that this particular sort of 'hearing' is always initiated by God.

However, such 'spiritual understanding' is not received without initial 'repentance of one's own sins' and initial 'faith' in Christ as Savior and Lord according to the 'conver(sion)' that is a promised result as well in Mark 4: 12 for the condition of "seeing" (perceiving) and "hearing" (understanding). Finally, it should be noted that the axiom of 'applying God's Standard' to oneself assumes that one 'perceives', 'understands', and is 'obeying' what God's Standards are and in this light, 'study' of the Holy Bible is seen to be implied by 'applying God's Standard' to oneself.

17) Why is 'receiving God's Grace' not an axiom since the grace of God is received in salvation when one expresses sincere 'faith' in Christ? The words, 'grace' and 'mercy' (which are different than 'agape' love as addressed in 8)

above), and its other forms as adjectives or verbs are not in the words of the conditional/procedural statements attributed to Christ in the Holy Bible. Also, none of the synonyms for 'grace' or 'mercy' are found in the conditional/procedural statements attributed to Christ in the Holy Bible. There are also no conditions in the conditional/procedural statements attributed to Christ in the Holy Bible that address the concepts associated with 'grace'; however, there is one (1) instance in which 'merciful' is noted as a positive condition (see Matthew 5: 7). Notably though, in Luke 6: 36, a believer's being 'merciful' is there conditioned upon recognition of God's being "kind unto the unthankful and evil". So, the greater antecedent for 'merciful' is recognition of God's Benevolence to those undeserving of His Mercy, as addressed in 13) above.

18) Why is 'knowing the will of God' not an axiom? Christ only spoke of 'doing the will of God', not 'knowing the will of God', and in so doing He addressed 'doing' the will of God' as a condition in all cases in the conditional statements attributed to Christ in the Holy Bible. Also, a 'willingness' to 'do the will the God' precedes the actual 'doing of His Will'. Moreover, in Luke 8: 10, Christ conditioned 'know(ing) the mysteries of the kingdom of God' upon being a disciple (which is already addressed in 2) above). Thus, 'knowing the will of God' is a result of remaining a true disciple of Christ, which involves all of the 'axioms' stated above, especially including persistence.

19) Why are the other 'fruits of the Spirit', besides 'agape love' as addressed in 8) and 'patience' in 15) above, not part of the axioms? The Holy Spirit was not given as a general gift to believers while Christ was on the earth, which Christ said in John 16: 7, and the 'fruits of the Spirit' cannot be realized without a believer having the Holy Spirit given to him/her. Thus, the other 'fruits of the Spirit' (joy, peace, gentleness, goodness, faithfulness, meekness, and temperance) are all seen as 'outgrowths' or 'extensions' of a life dedicated to following these axioms and the work of the Holy Spirit in such a believers.

20) Why are the six (6) parts of 'the armor of God' not axioms? The six (6) parts of the 'armor of God' are: 'loins girt with truth'; 'the breastplate of righteousness'; 'feet shod with the preparation of the gospel of peace'; 'the shield of 'faith''; 'the helmet of salvation'; and 'the sword of the Spirit, which is the word of God'? The 'truth' and 'the word of God' are stated literally by Christ to be logically equivalent in John 17: 17, and 'study' of 'the word of God' is addressed in 16) above. All believers receive the 'righteousness of Christ' upon expressing 'faith' in Christ, which is an 'axiom', but 'personal righteousness' is addressed in 2) above. All believers have the gospel of peace by 'faith', by receiving Christ (who said He gives His Peace to disciples as per John 14: 27), and by the 'fruit of the Spirit' for peace as addressed in 19) above, though it is important to note that the idea of 'having one's feet shod

with the preparation of the gospel of peace' is understood also to have the concept of 'witnessing', which is an axiom. Finally, all believers have salvation.

21) Why is receipt of certain 'spiritual gifts' not an axiom? Spiritual gifts are always a result of one believing in Christ and hence of receiving the Holy Spirit, from whom all spiritual gifts flow to a particular believer. Christ did not speak of 'spiritual gifts' explicitly as such in the Holy Bible, though he did state that to the disciples was given the gift to 'know the mysteries of the kingdom of God', which is addressed in 18) above. He also spoke in Matthew 25: 14-30 at length and pointedly in Matthew 13: 12 regarding differences in 'talents' (including 'gifts from God'), which all point to the axiom of 'sacrificial use of one's time, talent, and resources'. Additionally, in Matthew 7: 11, Christ also conditioned receipt of 'gifts' from the Father upon 'ask(ing)', which points to 'prayer' as addressed in 1) above. Next, Christ told the disciples in Matthew 10: 19 that God would give them the words to speak when they were being persecuted, which points directly to 'discipleship' as addressed in 2) above. Christ also said that if anyone would come unto Him, then He 'will give them rest', which assumes that the person first expresses 'faith' and then once a believer, comes to God the Father in 'prayer' as addressed in 2) above. Further, 'hear(ing)' the word of God and Christ is also associated with increased, spiritual blessing in Mark 4: 24, but this points back to 16) above. Very importantly, Christ also said that 'bring(ing)' forth fruit' is a key condition to successful prayer, but 'bring(ing) forth fruit' refers to 'personal righteousness' and to "making disciples", which are addressed in 2) above. Miraculously through God's grace and Christ's sacrifice, in John 17: 24, Christ says that he glorifies his disciples (of all ages within the context of John 14-17) with "the glory which thou (God the Father) gavest me (Jesus)", but this is not conditioned beyond being a disciple as addressed in 2) above. Finally, 'sacrificial' giving is the condition in Luke 6: 38 for one to receive 'gifts' from God {i.e., 'give' to 'be given' (from God)}, which results from the 'axiom' of 'sacrificial use of one's time, talent, and resources'. There is though a key point that must be addressed here in that Christ did make allowance for certain individuals to receive the 'gift' of celibacy in Matthew 19: 11, but since Christ stated clearly that this is not a type of 'gift' available to all believers, it is not an 'axiom' to be generally received nor in any case to be pursued.

22) Is there an order of importance for the twelve (12) axioms? Christ did not explicitly state any priorities in the Holy Bible regarding these attitudes and axioms. Rather, He presented these characteristics of a disciple as an integrated whole. Having said that, there is a practical, albeit imperfect, way to consider prioritization for the axioms. Specifically, the frequency with which

each axiom is present in Christ's statements in the Holy Bible is one such way to non-authoritatively ascribe 'importance' or 'priority' for these 'axioms'. As such, Christ had the most to say about 'sacrificial use of one's talents, time, and resources' as a condition to other promises. Indeed, such 'sacrifices' imply more positive conclusions in Christ's Words than any other 'axiom'. This is not meant in any way to diminish the other axioms. The ordering from 1) to 12) of the twelve (12) 'axioms' as listed at the beginning of this chapter corresponds to the frequency with which Christ employed these conditions, though it should definitely note be thought that 'witnessing' or 'praising God' are significantly less important to Christ than the other 'axioms'. Indeed, Christ's last words while on the earth pertained to 'witnessing' and "making disciples", and also without repentance and forgiveness, Christ's statements are clear that no prayer can be successful.

23) What disciple of Christ can keep all of these 'axioms' as this seems to be a rather strenuous list of disciplines to which a disciple must adhere in order to receive the promises in Christ's Words in the Holy Bible ? All disciples can keep these 'axioms', though only within the enabling power of the Holy Spirit and with single-minded a persistent, sacrificial commitment to Christ. Oswald Chambers called such commitment, "abandonment to God" in "My Utmost for His Highest". Christ spoke plainly about this matter when he soberly stated in Matthew 7: 14 that "narrow is the way that leadeth unto life, and few there be that find it", where in the original Greek wording, "few" means effectively "a very small, 'puny' group". Moreover, Christ is to have stated in John 17: 13 that "the Comforter (the Holy Spirit) shall guide you into all truths, of which these 'axioms' are considered a key part.

24) Why is 'forgiveness' its own separate axiom? Forgiveness by a believer of others' sins against that particular believer is explicitly required as a condition in Matthew 6: 14-15 for receiving forgiveness from God for a believer's own sins and also importantly for the 'inverse' statement as well in which the lack of a believer forgiving others' sins against that particular believer results in God's Lack of Forgiveness toward that believer). In turn, forgiveness of others' sins and hence the forgiving believer receiving forgiveness from God for their own sins is vitally connected to the 'axiom' for 'repentance' of one's own sins since Matthew 6: 15 implies that no 'repentance' can be achieved if such forgiveness is missing and hence in that case that none of the conclusions/promises associated with 'repentance' such as effective 'pray(ing)', can be received by a believer. However, it is not the case that 'forgiveness' of others' sins implies 'forgiveness of *all* of one's own sins from God' necessarily since there may be and likely are in many cases other sins on the part of a believer besides the lack of forgiveness of other's sins which will also require 'repentance'. Thus, 'repentance' of a more general

type beyond just 'forgiveness of other's sins' against a believer is also required in the life of a true disciple of Christ since such 'repentance' also involves one's own spiritual actions, independent of any other person's actions. Moreover, the other 'axioms' from this list and the conclusions implied by these 'axioms' in the statements attributed to Christ in the Holy Bible do not imply directly or indirectly any conclusion assuring the believer that this very necessary 'forgiveness of others' sins' can be assured other than via the direct act of 'forgiv(ing) all persons of all sins' against a believer.

25) Why is 'repentance of one's own sins' an axiom? Christ specifically stated in Luke 13: 5 that "not perishing" is conditioned explicitly upon "repent(ance)", and in Luke 15: 7 that "there is joy in heaven" when a sinner "repenteth". It should be noted though in Mark 4: 12 that Christ stated two (2) human conditions that precede 'repentance'—specifically, 'hear' (in the sense of understand spiritual truth about Christ) and 'see' (in the sense of perceive spiritual truth about Christ), but such brings us directly back to the issues associated with 'hearing' addressed in 16) above that ultimately point to initial 'repentance' and 'faith' in Christ. More generally, there are no other conditions involving actions or attitudes to be taken by human beings, besides the type addressed in 24) for 'forgiveness', that imply 'repent(ance) of one's own sins'.

26) Why is 'trust' (as expressed by total confidence in Christ and as distinguished from 'faith') not an 'axiom'? The direct answer to this question is that no instances of trust ('absolute confidence' in Christ) are found in the conditional statements attributed to Christ in the Holy Bible, though the concepts of trust are found in declarative statements from Luke 11: 2b and 3 ("Thy will be done, as in heaven, so in earth. Give us day by day our daily bread."). Indeed, all New Testament Greek words related to trust in any manner have been researched, but no such occurrences have been found. This is definitely the most surprising result to this author from this research since 'trust' as "absolute confidence' in God and Christ is such a significant part of the commanded and commended parts of a Christian believer's life in the New Testament overall. Such 'trust' assumes 'faith', so in that sense is such that 'faith' occurs in time before this time of 'trust' (absolute confidence). The two (2) instances of 'trust' being mentioned in the words attributed to Christ in the Holy Bible are noted in a 'negative' context, that is Christ stating in what one should *not trust* (i.e., see "trust in riches" via Mark 10: 24 and "Moses . . . in whom ye trust . . ." via John 5: 45). Thus, no argument for 'trust in the sense of 'absolute confidence' can be supported from this analysis.

27) Why is 'peacemaker' an 'axiom'? In Matthew 5: 9, Christ said that "blessed are the peacemakers: for they shall be called the children of God",

and there is one (1) other set of statements attributed to Christ that directly pertain to the concept of 'peacemak(ing)' (see Matthew 18: 15-17, where Christ gave specific instructions for addressing 'conflicts' or "trespasses" within the church). So, one who makes peace as a "peacemaker", as opposed to one who has 'peace' from being "in (Christ)" as discussed in 18) and 19) above, has no other conclusions to which it is attached than the two (2) in Matthew 5: 9 and Matthew 18: 125-17, and also "peacemaker" is not implied by any other condition. 'Making disciples' and all other 'axioms' are distinguished from 'peacemak(ing)' in that there is no indication in Matthew 5: 9 of 'peacemak(ing)' having any particular goal other than 'peace' with or among others and in Matthew 18: 15-17 of 'peacemak(ing)' having only the goal of restoring fellowship between believers. Also, the Greek word translated as "peacemaker" only has one (1) generally accepted meaning of one who is a 'maker of peace', though there are some who view "peacemaker" as also meaning 'a lover of peace'. Lastly, since 'peacemak(ing)' is directly associated with being a 'child of God' and is not implied by other conditions, the condition of being a "peacemaker" is considered fundamental.

28) Why is 'service to Christ' not its own axiom? While it is definitely true that 'service' is mentioned in several different settings by Christ in talking to the disciples, 'humility' and 'sacrificial use' are the primary 'axioms' assumed for both the attitude and accompanying actions for 'service' towards Christ and other believers to occur, such as in Matthew 25: 34-36 and in Matthew 20: 26-27, with 'faith' and 'desire' again assumed in all instances.

29) Why is 'do not fear' or 'do not be afraid' an 'axiom'? Matthew 14: 22-33 indicates that there is an inversely proportional relationship between 'true faith' and 'fear', where 'faith' is assumed to grow as in this passage. So, a growing 'faith' already addresses 'do not fear'. (NOTE: This type of 'fear' is that which is not of faith, while 'reverencing' God and Christ is a faithful act which can engender other positive conclusions.)

30) Why is 'do not seek to be rich' an 'axiom'? In Matthew 19: 16-26, which relates the real issue for the rich, young ruler to be lack of 'desire' and unwillingness to 'sacrifice', and in Matthew 6: 19-21 and in Luke 12: 16-21, both of which correlate one's 'heart' to where one's "treasure is" via the 'axioms' of 'desire' and 'sacrifice', Christ points to greater antecedents than "desire for riches" alone.

31) Why is 'do not worry' an 'axiom'? Matthew 6: 22-34 points directly to 'true desire' and 'faith' as the prerequisites for 'not worry(ing)'. Note especially verses 24-25 in this regard.

32) Why is 'do not remove offences' an 'axiom'? 'Repentance' is the condition for those with 'faith' and 'desire' that seeks to remove such "offences" from the life a disciple of Christ.

33) Why is 'do not have pride' an axiom? Matthew 5: 3, 5 and Luke 18: 9-14 all present 'humility' as a fundamental virtue, and where 'true humility' continues to exist, problems with 'selfish human pride' will not be consistently noted.

34) Why is 'do not offend' an axiom? 'Desire', 'humility', and 'applying God's Standards to others and self' are the positive conditions to address 'not offend(ing)'.

35) Why is 'Wisdom' not an 'axiom'? 'Wisdom' is given by God to those who have 'faith', 'desire', and 'humility' as per Matthew 11: 25, so there are other existing 'axioms' that imply 'wisdom'.

36) Why is 'meditate' not an 'axiom'? There is no indication of 'meditation' or its close synonyms or directly related concepts in the words attributed to Christ. In fact, statements such as those in Mark 13: 11 and Matthew 6: 34 indicate that forethought as to what to say or what will happen next (tomorrow) is not a part of the disciple's life. 'Meditation' is more than forethought though, but nonetheless, while present in other parts of the Scripture, there is no indication of such as a command or condition in the words accorded to Christ.

Final note: The exclusion of a particular 'good' action or attitude, such as prayer, love, personal righteousness, and Bible study, as an axiom is no manner intended to discourage these activities. Indeed, it is the essential position of these analyses that all of these actions and others noted above are needful, but that there are more fundamental, 'unimplied' actions and attitudes, which if cultivated carefully with God's Help, will serve to produce, again by God's Promises in Christ's Words and by His Grace and Help, these 'good' actions. Thus, this work attempts not to downplay the importance of such Christian activities and inner workings as these 'good' things, but rather to distill what are the 'best' things which will lead, according to the words of Christ in the Holy Bible, to one also doing and/or being the 'good' things or receiving spiritual aid from God.

Appendix A

Annotated List of All Conditional/Procedural Forms from Words Attributed to Jesus Christ in the Four (4) Gospels, Including Contexts and Cross-Referencing For All Forms

CONDITIONAL/PROCEDURAL FORMS IN MATTHEW

1. 4:19- IF follow me, THEN I will make you fishers of men. (MAGR-F) (I) (Disc) (Inductive)
Contrapositive: verse?, # 394 in Mark

2. IF you are poor in spirit, then yours is the kingdom of heaven. (MAGR-N) (I) (OT-R IS 57:15) (Disc)
Contrapositive: (verse?) ACT OR ATTITUDE? BOTH—see # 608

3. IF you mourn, THEN you shall be comforted. (MAGR-F) (I) (OT-Is 61:1–2) (Disc) (Inductive)
Contrapositive: verse? ACT OR ATTITUDE? BOTH

4. IF you are meek, THEN you shall inherit the earth. (MAGR-F) (I) (OT-Ps 37:11) (Disc) (Inductive)
Contrapositive: verse? ACT OR ATTITUDE? BOTH

5. IF you hunger and thirst after righteousness, then you shall be filled. (MAGR-F) (I) (OT-R Is 55:1–3) (Disc) (Inductive) ACT OR ATTITUDE OF HUNGER? BOTH
Contrapositive: verse?—see # 609 for similar (without "and thirst for righteousness")

6. If you are merciful, then you will obtain mercy. (MAGR-F) (I) (Disc) (Inductive)

Contrapositive: (verse?) IS MERCIFUL AN ACT OR SET OF ACTS OR AN ATTITUDE? I FAVOR BOTH, WHICH MEANS THIS IS NOT JUST A DESIRE FOR SHOWING MERCY BUT A HABIT OF LIFE AS ANACTION.

7. IF you are pure in heart, then you will see God. (MAGR-F) (I) (Disc) (Inductive)
Contrapositive: verse? ACT OR ATTITUDE OF PURE? BOTH

8. IF you are a peacemaker, then you shall be called a child of God. (MAGR-F) (I) (Disc) (Inductive)
Contrapositive: verse? ACT OR ATTITUDE OF PEACEMAKING? BOTH

9. IF you are persecuted for righteousness sake, then yours is the kingdom of heaven. (MAGR-N) (I) (Disc)
Contrapositive: verse?

10. If you are persecute(d) and (have men to) say all manner of evil against you for my sake, (then) Rejoice, and be exceeding glad. (MAGR-N) (I) (Disc)
Contrapositive: verse?—NOT in Mark, Luke, or John

11. If the salt has lost his savour, then it (salt) is good for nothing. (Disc) (Analogy) (MAMR)
Contrapositive: verse?—see # 492 in Mark for an inverse-like passage (though not strictly)—see # 842 for same

12. If a city is set on a hill, then it (the city) cannot be hid. (Disc) (I)(Analogy-literal) (MAMR)
Contrapositive: verse?

13. If you put a light on a candlestick, (then) it will give light to all in the house. (Disc) (I) (Analogy-literal) (MAMR) (Inductive)
Contrapositive: verseF?

14. If you let your light shine, then they (people) may see your good works and (may) glorify your Father who is in heaven. (Disc) (I) (Analogy-symbolic) (MAGR-N)
Contrapositive: verse?

{
15. If you break one of these least commandments and shall teach men so, then you shall be called the least in the kingdom of heaven. (Disc) (I) (MAGR-F)
Contrapositive: verse?

16. If you do and teach them (the least of these commandments), then you shall be great in the kingdom (Disc) (I) (MAGR-F) (Inductive) (STINV- close but technically needs "do or teach" for human logic)
Contrapositive: verse? }- see # 511 and 512, 224, 225 for different condition, but similar conclusion and also see # 180

17. If your righteousness does not exceed the righteousness of the scribes and Pharisees, then you shall in no case enter into the kingdom of heaven. (Disc) (I) (MAGR-F)
Contrapositive: verse?

18. If you kill, then you shall be in danger of the judgement.(Disc) (I) (OT-R) (MAGR-F) ? Note: since this may be a human court of judgement, then this may not be God's Judgement.
Contrapositive: verseF?(if human), but if God's judgement, then verse?

19. If you are angry with your brother without a cause, then you shall be in danger of the judgement. (Disc) (I) (MAGR-F)? See note on #18 for same.
Contrapositive: verseF? ?(if human), but if God's judgement, then verse?

20. If you say (to your brother) Raca (vain fellow), then you shall be in danger of the council. (Disc) (I) (MAMR)
Contrapositive: verseF?

21. If you say (to your brother) thou fool, you shall be in danger of hell fire. (Disc) (I) (MAGR-F)
Contrapositive: verse?

22. If thou bring thy gift to the altar and there remember that thy brother has aught against thee, then leave thy gift and be reconciled with thy brother. (Disc) (MAMR)
Contrapositive: verseF?

23. If thou bring thy gift to the altar and there remember that thy brother has aught against thee, and leave thy gift and be reconciled with thy brother, then come and offer thy gift. (Disc) (I) (MAMR)
Contrapositive: verse?

24. If you are in the way with your (enemy), then agree with him quickly. (Disc) (I) (MAMR)
Contrapositive: verseF? (1 person**********)

25. If you do not agree while you are in the way with your enemy, then you will be delivered to the judge and then to the officer and then to prison.
Paraphrase (Disc) (I) MAMR)
Contrapositive: verseF?

26. If you are thrown into prison, then you shall not come out (from prison) until you have paid the utmost farthing. (Disc) (I) (MAMR)
Contrapositive: verse? (1 person************)

27. If you look on a woman to lust after her, then you have already committed adultery with her in your heart. (Disc) (I) (OT-R) (MAMR)
Contrapositive: verse?

28. If your right eye offend you, then pluck it out and cast it from you. (Disc) (Analogy) (MAMR)
Contrapositive: verseF? (1 person**********)—command of Jesus

29. If your right hand offend you, then pluck it out and cast it from you. (Disc) (Analogy) (MAMR)
Contrapositive: verseF? (1 person**********)—command of Jesus

30. (If one) puts away his wife, (then) let him give her a writing of divorcement. (Disc) (I) (MAMR)
Contrapositive: verse?

31. (If one) put away his wife, except for the cause of fornication, (then that one) causeth her to commit adultery. (Disc) (I) (MAMR)
Contrapositive: verseF? (2 persons)—see # 201 for same and # 497 for similar (the "except" clause is not in Mark- # 497)

32. If one marries a woman that is divorced, (then this one) committeth adultery. (Disc) (I) (MAMR)
Contrapositive: verse? (1 person)

33. If (anything in communication) is more than Yea, yea, Nay, Nay, then (that thing) cometh of evil. (Disc) (I) (MAMR)
Contrapositive: verse?

34. If one shall smite you on your right cheek, then turn to him the other (cheek) also. (Disc) (I) (Analogy) (OT-R) (MAMR)
Contrapositive: verseF? (2 diff. persons)—see # 624

35. If any man will sue thee at the law, and take thy coat, then let him have thy cloak also. (Disc) (MAMR) (2 diff. persons
Contrapositive: verseF?—see # 625 for very similar

36. If someone shall compel you to go a mile, then go with him twain. (Disc) (I) (MAMR)
Contrapositive: verseF?

37. If (one) asketh you, then give to him. (Disc) (I) (MAMR)
Contrapositive: verseF?—see # 627 for same essentially

38. If (one) would borrow of you, then turn him not away. (Disc) (I) (OT-R, close to quoting) (MAMR)
Contrapositive: verseF?

39. If one is your neighbor, then love him. (Disc) (I) (OT-Q) (MAMR)
Contrapositive: verseF? (NOTE: Matt 5: 43b is not addressed here as this appears to be Jesus using words of the religious leaders and is not quoting Scripture.)

39a. If one is your enemy, then love them. (Disc) (I) (MAMR)
Contrapositive: verseF? (2 persons)—see # 620 for same and # 631 for same again

40. If one curse you, then bless them. (Disc) (I) (MAMR)
Contrapositive: verseF?—see # 622 for same

41. If one hates you, then do good to them (those who hate you). (Disc) (I) (MAMR)
Contrapositive: verseF? (2 different persons)—see # 621

42. If one despitefully uses you and persecutes you, then pray for them. (Disc) (I) (MAMR)
Contrapositive: verseF? (2 persons)—see # 623 for same

43. If you love your enemies, bless them that curse you, do good to them that hate you, and pray for them that despitefully use you and persecute you, then you may be the children of your Father which is in heaven. (Disc) (I) (MAGR-F)
Contrapositive: verse?

44. If you do your alms before men to be seen of them, then you have no reward of your Father which is in heaven. (Disc) (I) (MAGR-N and F)
Contrapositive: verse?

{
45. If you forgive men their trespasses, your Heavenly Father will also forgive your trespasses. (Disc) (MAGR-N and F) (Inductive)
Contrapositive: verse?—see # 637 for very similar

46. If you do not forgive men their trespasses, then neither will your Father forgive your trespasses. (Disc) (MAGR-N and F) (Inductive)- Contrapositive: verseF?} (STINV)- MAGR—SEE # 416 for another condition that implies God will not forgive (the sin against the Holy Spirit)

46a) . . . when you fast, (then ye) be not as the hypocrites, of a sad countenance . . . (Disc) (MAMR) (INFERRED)
Contrapositive: verseF?

46b) . . . . (since) they (the hypocrites) disfigure their faces, that they may appear unto men to fast, . . . (then ye) be not, as the hypocrites, of a sad countenance . . . (Disc) (MAMR) (INFERRED)
Contrapositive: verseF? (what one man does would not seem necessarily to cause directly another man to do something or not)

46c) . . . when thou fastest, (then thou) anoint thine head, and wash thine face . . . (Disc) (MAMR) (INFERRED)
Contrapositive: verseF?

46d) . . . (if) thou appear not unto men to fast, but unto thy Father which is in secret . . . (then) thy Father which seeth in secret, shall reward thee openly. (Disc) (MAGR) (INFERRED) (Inductive)
Contrapositive: verse?

}
47. If you lay up treasures for yourselves upon earth, then moth and rust doth corrupt (them) and thieves break through and steal (them). (Disc) (I) (OT-R) (MAMR) (2 persons)
Contrapositive: verse?

48. If you lay up treasures for yourselves in heaven, then neither moth nor rust doth corrupt (them) and thieves do not break through and steal (them). (MAMR)
Contrapositive: verse? } (close STINV)

49. If your treasure is (somewhere), then there your heart will be also. (Disc) (I) (MAMR) (Inductive)
Contrapositive: verse?—same as # 786

{
50. If your eye is single, then your whole body shall be full of light. (Disc) (Analogy-symbolic) (MAGR-N&F) (Inductive)
Contrapositive: verse?—see # 754 for same conclusion, but different condition

51. If your eye is evil, then your whole body shall be full of darkness. (Disc) (Analogy) (MAGR) (Inductive)
Contrapositive: verse? } (STINV)

52. If the light that is in you be darkness, then how great is that darkness. (Disc) (Analogy) (MAMR)
Contrapositive: verse?

53. If God so clothe the grass of the field (which is arrayed in more glory than Solomon), which to day is, and to morrow is cast into the oven, then shall he not much more clothe you. (Disc) (I) (GAGR-N)
Contrapositive: verseF?

*** 54a. If you seek first the kingdom of God and his righteousness, then all these things shall be added unto you. (Disc) (I) (OT-R) (MAGR-F) (Inductive)
Contrapositive: verse?—could be instance where MAGR is false in contrapositive? But, these things (eat, drink, clothing) are needs not wants. Psalms says that "I have never seen the righteous begging for bread." (see # 57 and #783 for same)

54. If you judge (with a certain type of judgement), then you shall be judged (by that same judgement. (Disc) (I) (MAGR-F) (Inductive)
Contrapositive: verse?—see # 635 for very similar

55. If you first cast out the beam out of your own eye, then you shall see clearly to cast out the mote out of your brother's eye. (Disc) (I) (MAMR) (Inductive)
Contrapositive: verse?—see # 643 for same

56. If you cast your pearls before swine, then they trample under their feet and turn again and rend you. (Disc) (I) (Analogy) (OT-R) (MA(MR?), not MAGR)
Contrapositive: verse?

57. ***If you ask, (then) it shall be given unto you. (Disc) (I) (MAGR-N&F) (Inductive) THIS IS THE 1st INSTANCE IN STUDY WHERE MAGR HAS CONTRPOSITIVE IS FALSE. (see 54)
*Contrapositive: verseF? Paul's case of thorn in flesh- II Corinthians—see # 731 for same essentially

58. ***If you seek, then you shall find. (Disc) (I) (MAGR-N&F) (Inductive)
Contrapositive: verseF? Paul???—see # 732 for same

59. *** If you knock, then it shall be opened to you (Disc) (I) (Analogy) (MAGR-N&F) (Inductive)
*Contrapositive: verseF? Paul in Acts with desire to go to Asia, but lead otherwise. THIS IS THE 3rd INSTANCE IN STUDY WHERE MAGR HAS CONTRPOSITIVE IS FALSE. (see # 54, # 57)—see # 733 for same

60. *** If one asketh, then (one) receiveth. (Disc) (I) (MAGR-N&F) (OT-Q)
Contrapositive: verseF? Paul – thorn of the flesh . THIS IS THE 4th INSTANCE IN STUDY WHERE MAGR HAS CONTRPOSITIVE IS FALSE. (see 54, 57, 58)—see # 233 and # 1348 for very similar, but not the same

61. If (one) seeketh, (one) findeth. (Disc) (I) (MAGR-N&F) (OT-Q)
Contrapositive: verseF? Paul??

62. If (one) knocketh, then it shall be opened. (Disc) (I) (MAGR-N&F) (Analogy) (Inductive)
Contrapositive: verseF- Paul –thorn in the flesh THIS IS THE 5th INSTANCE IN STUDY WHERE MAGR HAS CONTRPOSITIVE IS FALSE. (see 54, 57, 58, 59)

63. If you then, being evil, know how to give good gifts unto your children, then how much more shall your Father which is in heaven give good things to them that ask him? (Disc) (I) (MAGR)
Contrapositive: verseF?

64. If something is what you would that men should do to you, then you do so (the same thing) to them. (Disc) (I) (MAMR)—See # 628 for same
Contrapositive: verse?
{
65. If a tree is good, then (the tree) bringeth forth good fruit. (Disc) (I) (Analogy-symbolic) (None)
*Contrapositive: verse—see # 644 see and also # 69 below

66. If a tree is corrupt, then (the tree) bringeth forth evil fruit. (Disc) (I) (Analogy-symbolic) (None)
Contrapositive: verse?—see # 645
} CLOSE STINV

{
67. If a tree is good, then (the tree) cannot bring forth evil fruit. (Disc) (I) (Analogy-literal) (None)
Contrapositive: verseF?—see # 644 for same

68. If a tree is corrupt, then (the tree) cannot bring forth good fruit. (Disc) (I) (Analogy-literal) (None)
Contrapositive: verseF?—see # 645 for same
} Close STINV

69. If a tree bringeth not forth good fruit, then (the tree) is hewn down and cast into the fire. (Disc) (I) (Analogy-literal) (None)
Contrapositive: verse?—see # 813 for very similar (essentially the same, though the casting into the fire is not in #813

70. If one does the will of my Father which is in my heaven, (one) shall enter into the kingdom of heaven. (Disc) (I) (MAGR-F)
Contrapositive: verse? (see # 417 for same condition, but different conclusion slightly (A—>B)

71. If one says to me (Jesus), Lord, Lord, but doeth not the will of my Father which is in heaven then (that one) shall not enter heaven . . . and (then) will I (Jesus Christ) profess unto them, I never knew you. (Disc) (I) (OT-R) (MAGR-F)
Contrapositive: verse?

{
72. If (one) heareth these sayings of mine, and doeth them, then I will liken him unto a wise man, which built his house upon a rock. (Disc) (I) (Analogy) (MAGR-F) (Inductive)
Contrapositive: verse?—see # 651 for same, see # 686 for same condition, but different conclusion

73. If (one) heareth these sayings of mine, and doeth them not, then he shall be likened unto a foolish man, which built his house upon the sand. (Disc) (I) (Analogy) (MAGR-F) (Inductive)
Contrapositive: verse? } (close STINV) – not true inverse, but close

74. If (one) is a child of the kingdom (of the children of Abraham, Isaac, and Jacob at the time of the kingdom of heaven), then (one) shall be cast out into outer darkness. (Mult) (I) (MAGR-F)
Contrapositive: verse?

{
75. If (one) is whole, then (one) need(s) not a physician. (Mult) (I) (Analogy-literal) (MAMR)
Contrapositive: verse?—see 398 & 598

76. If (one) is sick, then (one) need(s) a physician. (Mult) (I) (Analogy-literal) (MAMR)
Contrapositive: verse? } (close STINV) – not true inverse but close

76 a) . . . when the bridegroom shall be taken from them (the children of the bridechamber), . . . then shall they (the children of the bridechamber) fast. (Disciples of John the Baptist) (GAMR)
Contrapositive: verse?—see # 401 for same and # 400 for inverse

77. If a man putteth a piece of new cloth unto an old garment, then that which is put in to fill it up taketh from the garment and the rent is made worse. (Mult) (I) (Analogy-literal) (MAMR)
Contrapositive: verseF?

77a)—see # 600

{
78. If men put new wine into old bottles, then the bottles break and the wine runneth out and the bottles perish. (Mult) (I) (Analogy-symbolic & literal) (MAMR)
Contrapositive: verseF?

78a) see # 602

79. If men put new wine into new bottles, then both (the wine and the bottle) are preserved. (Mult) (I) (Analogy-literal) (MAMR)
Contrapositive: verseF? } (STINV) – close but not exact

80. If you have received freely, then freely give. (Disc) (I) (GAMR)
Contrapositive: verseF?

81. (If you enter) into (a) city or town, then (you shall) inquire who in it is worthy and there abide until you go thence. (Disc) (I) (MAMR)
Contrapositive: verseF?

{
82. If the house be worthy, (then) let your peace come upon it. (Disc) (MAMR)
Contrapositive: verseF?

83. If it (the house) be not worthy, let your peace return to you. (Disc) (MAMR)
Contrapositive: verseF? } (STINV)

84. If (one) does not receive you, nor hear your words, when you depart out of that city, (then) shake off the dust of your feet. (Disc) (I) (MAMR)
Contrapositive: verseF? (2 person)—see # 96 for opposite of condition, and different conclusion than here (but not inverse)—see # 689 for same

84a) . . . when they (men) deliver you up, (then) take no thought how or what ye shall speak . . . (Disc) (MAMR) (INFERRED)
Contrapositive: verseF?—see # 548 for same

84b) . . . (since) it shall be given you in that same hour what ye shall speak, (then) take no thought how or what ye shall speak . . . (Disc) (GAMR) (INFERRED)
Contrapositive: verseF?—see # 549 for same

84c) See # 550 for very similar.

84d) (If one) . . . endureth to the end, (then that one) shall be saved. (Inductive)
Contrapositive: -verse?—see # 294 & # 552 for same

84e) . . . when they (men) persecute you in this city, (then) flee ye into another (city) . . . (Disc) (MAMR) (INFERRED)
Contrapositive: verseF? – NOT in Mark

84f) . . . (since) verily I say unto, Ye shall not have gone over the cities of Israel, till the Son of man be come, (then) . . . when they (men) persecute you in this city, flee ye into another (city) . . . (Disc) (MA-GA/MR) (INFERRED)
Contrapositive: verse? –

84g) . . . till the Son of man be come, (then) verily I say unto, Ye shall not have gone over the cities of Israel, . . . (Disc) (GAMR) (INFERRED)
Contrapositive: verse? – NOT in Mark

85. If they have called the master of the house Beelzebub, (then) how much more (shall they call) them of his household? (Disc) (I) (MAMR) (Inductive)
Contrapositive: verseF?

86. If there is (something that) is covered, then (that something) shall be revealed. (Disc) (I) (Analogy-literal) (MAGR) (Inductive)
Contrapositive: verseF?—see # 765 for same

87. If (there is something that is) hid, then (that something) shall be known. (Disc) (I) (Analogy-literal) (MAGR)—See # 766 for same
Contrapositive: verseF?

{
88. If (ones are able to) kill the body but are not able to kill the soul, (then) fear not them. (Disc) (I) (MAMR)
Contrapositive: verseF? (2 person)

89. If (one) is able to destroy both soul and body in hell, then fear him. (Disc) (I) (MAMR)
Contrapositive: verseF? } STINV- close , but would be not body or the soul—See # 769 for essentially the same

Since the very hairs on your head are all numbered, then you should not fear.
{
90. If (one) shall confess me before men, (then) him will I confess before my Father which is in heaven. (Disc) (I) (MAGR-F)
Contrapositive: verse?

91. If (one) shall deny me before men, (then) him will I also deny before my Father which is in heaven. (Disc) (I) (MAGR-F)
Contrapositive: verse? } close STINV—but not confess seems not to be deny

92. If (one) loves father or mother, son, or daughter more than me, (then that one) is not worthy of me. (Disc) (I) (MAGR-N&F)
Contrapositive: verse? (One cannot be worthy of Jesus, so this is true always).

93. If (one) taketh not his cross and followeth after me, (then that one) is not worthy of me. (Disc) (I) (Analogy-literal) (MAGR-N&F)

Contrapositive: verse? (One cannot be worthy of Jesus, so this is true always). See #470 for similar condition (for the contrapositive of #470), but different conclusion.

{

94. If (one) findeth his life, (then that one) shall lose it. (Disc) (I) (MAGR-F) (Analogy) (Inductive)
Contrapositive: verse?

95. If (one) loseth his life for my sake, (then that one) shall find it. (Disc) (I) (MAGR-F) (Analogy) (Inductive)
Contrapositive: verse? } STINV

96. (If one) receiveth you (disciples), (then that one) receiveth me (Jesus). (Disc) (I) (MAMR)
Contrapositive: verseF? – See #479 for different condition, but same conclusion

97. If (one) receiveth me (Jesus), (then that one) receiveth him that sent me. (Disc) (I) (GA/GR-MR)
Contrapositive: verseF?- Jews receive God The Father, but reject Jesus. See # 480 for essentially the same—see # 1210 for essentially the same (though receive and believe on are not the same)

98. (If one) receiveth a prophet in the name of a prophet, (then that one) shall receive a prophet's reward. (Disc) (I) (MAMR) (Inductive)
Contrapositive: verse?—see # 613 for "inverse like"

99. (If one) receiveth a righteous man in the name of a righteous man, (then that one) shall receive a righteous man's reward. (Disc) (I) (MAMR) (Inductive)
Contrapositive: verse?

100. ***(If one) shall give to drink unto one of these little ones a cup of cold water only in the name of a disciple, (then that one) shall verily in no wise lose his reward. (Disc) (I) (MAGR-F)
Contrapositive: verseF? – (see verses 54, 57, etc. for MAGR with contra false?) (see (# 484 for very similar condition and same conclusion)

101. (If one) shall not be offended in me, (then) blessed is (that one). (Disc) (I) (OT-R) (MAGR-N&F)
Contrapositive: verseF?—see # 654 for same; see verses 54, 57, 99, etc. for MAGR with contra false?)

102. (If one) is least in the kingdom of heaven, (then that one) is greater than he (John The Baptist). (MULT) (I) (GA/MR-GR)
Contrapositive: verseF?—see # 657 for same; the converse also holds by equivalency inferred from definition from words attributed to Christ, ". . . he that is least in the kingdom of God is greater than he."

103. If you will receive it, (then) this (John The Baptist) is Elias, which was for to come. (MULT) (I) (OT-R) (MAGR)
Contrapositive: verseF?

104. (If one) hath ears to hear, (then) let (that one) hear. (MULT) (I) (MAMR)
Contrapositive: verseF?—see #459 and #845 for same

105. If the mighty works, which were done in you (Chorazin, Bethsaida), had been done in Tyre and Sidon, (then) they would have repented long ago in sackcloth and ashes. (MULT) (I) (OT-R) (GAMA)
Contrapositive: verseF?—see # 712 for same

106. If the mighty works, which were done in you (Capernaum), had been done in Sodom, (then) it (Sodom) would have remained until this day. (MULT) (I) (OT-R) (GAGR)
Contrapositive: verseF?

107. If you labor and are heavy laden and come unto me, (then) I will give you rest. (MULT) (I) (MAGR-N) (Analogy-literal) (Inductive)
Contrapositive: verse?

108. If you had known what this meaneth, I will have mercy and not sacrifice, (then) you would not have condemned the guiltless. (Relig) (I) (MAMR) (OT-Q)
Contrapositive: verse?

109. If it (a sheep) fall into a pit on the sabbath day, (then) will he not lay hold on it and lift it out? (Relig) (I) (MAMR) (OT-R)
Contrapositive: verseF?

110. If a kingdom is divided against itself, (then the kingdom) is brought to desolation. (Relig) (I) (MAGR) (Analogy-literal)
Contrapositive: verseF? – See # 734 for same

111. If a city or house is divided against itself, (then the city or house) shall not stand. (Relig) (I) (MAGR) (Analogy-literal) (Inductive)
Contrapositive: verseF?—see # 735 for same

112. If Satan cast out Satan, (then) he is divided against himself. (Relig) (I) (None)
Contrapositive: verseF?

113. Matt 12: 27—a therefore used, with preceding IF clause, but not clear as to Conditional/Procedural use.

114. If I (Jesus) cast out devils by the Spirit of God, then the kingdom of God is come unto you. (Relig) (GAGR)
Contrapositive: verseF?—see # 736 for very similar ("finger of God" used in # 736 instead of "the Spirit of God")

115. If one (does not) first bind the strong man, (then) one will not spoil the strong man's house. (Relig) (MAMR) (I) (Analogy-literal) (2 persons) (Inductive)
Contrapositive: verseF?—see # 737 for close to INVERSE

116. (If one) is not with me, (then one) is against me. (Relig) (MAMR) (I) (1 person)
Contrapositive: see # 739 for same; see #482 for very close to contrapositive in Mark 9:40 and see # 700 for very close to contrapositive also in Luke 9:50

117. (If one) gathereth not with me, (then that one) scattereth abroad. (Relig) (MAMR) (I) (Analogy-literal)
Contrapositive: verse?—see # 740 for essentially the same ("abroad" is not in #740)

118. (If one commits) blasphemy against the Holy Ghost, (then that one) shall not be forgiven. (Relig) (MAGR) (I)
Contrapositive: verseF?

119. (If one) speaketh a word against the Son of man, (then) it shall be forgiven him. (Relig) (MAGR) (I)—see item 57
Contrapositive: verseF?

120. (If one) speaketh against the Holy Ghost, (then) it shall not be forgiven him, neither in this world, neither in the world to come. (Relig) (MAGR) (I)
Contrapositive: verseF?—see item 57

121. (If one) shall speak an idle word, (then that one) shall give an account thereof in the day of judgment. (Relig) (MAGR) (I)—see item #57
Contrapositive: verseF?

122. (If) a generation is evil and adulterous, (then that generation) seeketh after a sign. (Relig) (MAMR) (I)—see item 57
Contrapositive: verseF? – Sodom

122a) When the unclean spirit is gone out of a man, (then) he walketh through dry places, seeking rest, and findeth none. (Relig) (unclean spirit – unclean spirit) (INFERRED)
Contrapositive: verseF? – NOT in Mark – See # 741 for same

122b) . . . when he (the unclean spirit) is come, (then) he findeth it (my house from whence I come) empty, swept, and garnished. (Relig) (unclean spirit – unclean spirit) (INFERRED)
Contrapositive: verseF? – NOT in Mark – See # 742 for same

122c) (when) he findeth it (my house from whence I come) empty, swept, and garnished Then goeth he, and taketh with himself seven other spirits more wicked than himself, and they enter in and dwell there: and the last state of that man is worse than the first. (Relig) (unclean spirit – unclean spirit) (INFERRED—"when" not given)
Contrapositive: verse? – NOT in Mark – See # 743 for same

123. (If one) shall do the will of my Father which is in heaven, (then that one) is my brother, and sister, and mother. (Mult) (MAGR) (I)
Contrapositive: verse?—converse holds because of definition from statement attributed to Christ as ". . . whosoever shall do the will of my Father which is in heaven, the same is my brother, and sister, and mother."

124. (If one) hath ears to hear, (then) let (that one- him) hear. (Mult) (I) (MAMR) (1 person)
Contrapositive: verseF?

{
125. (If one) hath, (then to that one) shall be given and (that one) shall have abundance. (Disc) (I) (MAGR-N&F) (Inductive)
Contrapositive: verseF?—salvation by grace—see # 350 for essentially the same

126. (If one) hath not, (then from that one) shall be taken even that he hath. (Disc) (I) (MAGR-N&F) (Inductive)
Contrapositive: verseF?—see item 57 and # 125 for opposites
} Close STINV

127. (When) any one heareth the word of the kingdom and understandeth it not, then cometh the wicked one, and catcheth away that which was sown in his heart. (Disc) (MA/Satan'sResult allowed by God)
Contrapositive: verseF?—see #426 and also see # 1140 to show that inverse need not be true as well

128. (If one) heareth the word, anon with joy receiveth it, (then that one) by and by is offended, who hath not root in himself, when tribulation or persecution ariseth because of the word. (Disc) (I) (MAMR?) (Analogy in part-symbolic) – See # 430
Contrapositive: verseF?—(See # 428 and # 429, # 430 for similar)

129. (If one) heareth the word; and the care of this world and the deceitfulness of riches, choke the word, (then that one) becometh unfruitful. (Disc) (I) (MAGR) (Analogy- in part)
Contrapositive: verse? (Note: assuming "not unfruitful" is not the same as "fruitful")

130. (If one) heareth the word, and understandeth it, (then that one) beareth fruit, and bringeth forth, some an hundredfold, some sixty, some thirty. (Disc) (I) (MAGR) (Analogy- in part - symbolic)
Contrapositive: verseF?—see # 425 for similar condition, but different conclusion

130a) . . . when the blade was sprung up, and brought forth fruit, then appeared the tares also. (Disc) (Analogy) (GAGR)
Contrapositive: verseF?—literal – NOT in Mark

130b) . . . when it (a mustard seed) is grown, (then) it (the mustard seed) is the greatest among herbs, and becometh a tree, so that birds of the air come and lodge in the branches thereof. (Disc) (GAGR) (Analogy—literal)
Contrapositive: verse?—see # 447 for essentially the same

131. (When) the end of the world (occurs), then the tares (children of the wicked one) are gathered and burned in the fire. (DISC) (I) (GAGR)
Contrapositive: verseF?

132. (When) The Son of man shall send forth his angels, (then) they (the angels) shall gather out of his kingdom all things that offend and them which do iniquity and shall cast them into a furnace of fire. (DISC) (I) (GAGR)
Contrapositive: verseF?

133. (When) (the angels) shall gather out of his kingdom all things that offend and them which do iniquity and shall cast them into a furnace of fire, (then) there shall be wailing and gnashing of teeth. (DISC) (I) (GAGR)
Contrapositive: verseF?—see #818 for similar

134. (When) (the angels) shall gather out of his kingdom all things that offend and them which do iniquity and shall cast them into a furnace of fire and there shall be wailing and gnashing of teeth, Then shall the righteous shine forth in the kingdom of their Father. (Disc) (GAGR)
Contrapositive: verseF?

135. (If one) hath hears to hear, (then) let him hear. (Disc) (I) (MAMR)
Contrapositive: verseF?

136. . . . When a man hath found (treasure in a field), (then) he hideth, and for joy thereof goeth and selleth all that he hath and buyeth that field. (Disc) (I) (MAMR) (Analogy-literal)
Contrapositive: verseF?—see # 501 for a somewhat converse-like passage in the symbolic sense of the "field" being as "heaven", but in this passage here the man "buyeth the field" so it is different as well

137. . . . When (a merchant man) had found one pearl of great price, (then he) went and sold all that he had, and bought it. (Disc) (I) (MAMR) (Analogy-literal)
Contrapositive: verseF?

138. . . . when it (a fishing net) was full, (then) they drew (it) to shore, and sat down and gathered the good into vessels, but cast the bad away. (Disc) (I) (MAMR) (Analogy-literal)
Contrapositive: verseF?

139. . . . (when) at the end of the world, (then) the angels shall come forth, and sever the wicked from among the just, And shall cast them into the furnace of fire: (then) there shall be wailing and gnashing of teeth. (Disc) (I) (GAGR)
Contrapositive: verse?

140. (If) a man is a scribe which is instructed unto the kingdom of heaven, (then that man) is like unto a man that is an householder, which bringeth forth out of his treasure things new and old. (Disc) (I) (MAMR-N) (Analogy-literal)
Contrapositive: verseF ?

141. (If one is) a prophet, (then that one) is not without honour, save in his own country, and in his own house. (Disc) (I) (GAMR)
Contrapositive: verse?

142. (If you keep) your tradition, (then) ye also transgress the commandment of God. (Relig) (I) (MAMR) (INFERRED)—Matthew 15:3
Contrapositive: verse?—see verse 4–6 of chapter. 15 in Matthew (same chap) for context on "tradition" as being a negative thing. See #455 for Converse (transgress is like reject)

143. (If one) curseth father or mother, (then) let him die the death. (Relig) (I) (MAMR-N)
Contrapositive: verseF?

144. (If) whosoever shall say to his father or his mother, It is a gift, by whatsoever thou mightest be profited by me and not honour his father or his mother, (then) he shall be free. (Relig) (I) (MAMR-F)
Contrapositive: verseF?

145. (If), by your tradition, (one says to his father or his mother, It is a gift, by whatsoever thou mightest be profited by me and not honour his father or his mother), thus have ye made the commandment of God of none effect. (Relig) (I) (MAMR-N)
Contrapositive: verseF? – See # 456 for very similar

{
146. (If a thing) goeth into the mouth, (then that thing does) not defileth a man. (Mult & Relig) (I) (MAGR-N&F) (Analogy-literal)
Contrapositive: verseF? (what about going into mind)—see #457 for essentially the same

147. (If a thing) cometh out of the mouth, (then that thing) defileth a man. (Mult & Relig) (I) (MAGR-N&F) (Analogy-literal)—see # 458 for essentially the same
Contrapositive: verseF?—(good things can come out of mouth)

} close STINV
NOTE: THIS is not considered a STINV strictly speaking since goeth and cometh are not negations necessarily.

148. (If any) plant hath not (been) planted (by) my heavenly Father, (then that plant) shall be rooted up. (Disc) (I) (GAGR) (Analogy-literal)
Contrapositive: verse?

149. If the blind lead the blind, (then) both shall fall into the ditch. (Disc) (MAMR) (Analogy-literal) (Inductive)
Contrapositive: verseF?

150. (If a thing) entereth in at the mouth, (then that thing) goeth into the belly, and is cast out into the draught. (Disc) (I) (MAMR)
Contrapositive: verse? – similar #461

151. (If) things proceed out of the mouth, (then those things) come forth from the heart. (Disc) (I) (MAMR)
Contrapositive: verseF? (man can speak with mind only)—see Matt 15: 19 in # 152

152. (If) things come forth from the heart, (then those things) defile a man. (Disc) (I) (MAGR-N)
Contrapositive: verseF? (bad thoughts can defile)—see Matt 15: 19 in # 147 and see # 458 for slightly different condition, but same conclusion

153. (If a thought or an action) is an evil thought, murder, adultery, fornication, theft, false witness, (or) blasphemies, (then that thought or action) proceed(s) from the heart. (Disc) (I) (MAMR)
Contrapositive: verseF?

154. (If a thought or an action) is an evil thought, murder, adultery, fornication, theft, false witness, (or) blasphemies, (then those thoughts or actions) are the things which defile a man. (Disc) (I) (MAMR?/MAGR)—see #458 for different condition but same conclusion
Contrapositive: verseF? (other sins by action?)

155. (If one) eat(s) with unwashen hands, (then this) defileth not a man. (Disc) (I) (MAMR?/MAGR)
Contrapositive: verseF?—see #146 and # 457 for very similar

156. (Because) they (the multitude) continue with me now three days, and have nothing to eat, I have compassion on the multitude. (Disc) (MAGR-N)
Contrapositive: verseF?—# 466 for same

157. (If) I (Jesus) send them (the multitudes) away fasting, (then) they (will) faint in the way. (Disc) (I) (GAMR)
Contrapositive: verseF?—#467 same

157a) . . . When it is evening, (then) ye say, It will be fair weather: for the sky is red. (Relig) (INFERRED) (GAMR)
Contrapositive: verseF? – could have bad weather and it still be evening. NOT in Mark

157b) (since) the sky is red (in the evening), (then) ye say, It will be fair weather . . . (Relig) (INFERRED) (GAMR)
Contrapositive: verseF? – men may not say things even though they are true—NOT in Mark

157c) . . . (since) the sky is red and lowring, . . . (then ye say) in the morning, it will be foul weather today . . . (Relig) (INFERRED) (GAMR)
Contrapositive: verseF?—men may not say things even though they are true - NOT in Mark

158. (If) a generation is wicked and adulterous, (then that generation) seeketh after a sign. (Relig) (I) (MAMR)
Contrapositive: verseF? (what of the heathen who seek after no sign?)

159. (If) a generation is wicked and adulterous, (then) there shall no sign be given it (that generation), but the sign of the prophet Jonas. (Relig) (I) (MAGR-N&F)
Contrapositive: verse?

160. (If or since) flesh and blood hath not revealed it unto thee, but my Father which is in heaven, (then) blessed art thou. (Disc) (I) (GAGR)
Contrapositive: verseF?

161. (If) thou shalt bind whatsoever on earth, (then that thing) shall be bound in heaven. (Disc) (I) (MAGR-F)
Contrapositive: verse?

162. (If) thou shalt loose whatsoever on earth, (then that thing) shall be loosed in heaven. (Disc) (I) (MAGR-F)
Contrapositive: verseF? (Satan?)

163. (If) thou savorest not the things that be of God, (then) thou art an offense unto me (Jesus). (Disc) (I) (MAMR)
Contrapositive: verse?

164. If any man will come after me, (then) let him and himself, and take up his cross, and follow me. (Disc) (MAMR)
Contrapositive: verse? (seems to be true vacuously)—see # 470 and # 690 for same

165. (If one) will save his life, (then that one) shall lose it (his life). (Disc) (I) (MAGR?)
Contrapositive: verseF? (can be saved without being martyr)—see # 94 for very similar though initial condition is slightly different and see # 471 for same

166. (If one) will lose his life for my sake, (then that one) shall find it (his life). (Disc) (I) (MAGR)
Contrapositive: verseF?—see # 95 for same and #472 for very similar

167. (When) the Son of man shall come in the glory of his Father with his angels, then he shall reward every man according to his works. (Disc) (I) (GAGR)
Contrapositive: verse? (statement is true vaccuously if condition is false).

168. (When) Elias truly shall first come, (then he shall) restore all things. (Disc) (I) (GAGR)
Contrapositive: verse? (statement is true vaccuously if condition is false).

169. If ye have 'faith' as a grain of mustard seed, (then) ye shall say unto this mountain, Remove hence to yonder place. (Disc) (MAGR-N) (Inductive)
Contrapositive: verseF?

170. If ye have 'faith' as a grain of mustard seed and ye shall say unto this mountain, Remove hence to yonder place, (then) it shall be removed. (Disc) (MAGR-N)
Contrapositive:

171. If ye have 'faith' as a grain of mustard seed, (then) nothing shall be impossible unto you. (Disc) (I) (MAGR) (Inductive)– see # 475 for similar condition and same basic conclusion
Contrapositive: verse?—see # 504 for similar in regard to contrapositive, but not exactly with wording

172. (If one has) this kind (of situation in which another one is devil-possessed), (then) this kind goeth not out but by prayer and fasting. (Disc) (I) (MAGR-N, with man's involvement)
Contrapositive: verse? (Note that STINV is implied as well but not explicitly stated)—See #476 for essentially the same

173. (When) The Son of man shall be betrayed into the hands of men, (then) they shall kill him. (Disc) (I) (MAMR) (2 persons)
Contrapositive: verseF?

174. (When) The Son of man shall be betrayed into the hands of men and they shall kill him, (then) the third day he shall be raised again. (Disc) (I) (MAGR-F)
Contrapositive: verse?—see # 477 for very similar

175. (If) the kings of the earth take custom or tribute of the strangers, then are the children free. (Disc-Peter) (I) (MAMR) (2 persons)
Contrapositive: verseF?

176. (When) thou hast opened his mouth (of the first fish you catch with a hook in the sea), (then) thou shalt find a piece of money. (Disc-Peter) (I) (MAGR-N)
Contrapositive: verse? (as a MAGR)

177. (When) thou hast opened his mouth (of the first fish you catch with a hook in the sea) and thou shalt find a piece of money, (then) take that and give unto them (tribute collectors) for me and thee. (Disc-Peter) (I) (MA-GA/MR)—1st part involves Peter too
Contrapositive: verseF?

178. (If) ye be converted, (then ye shall) become as little children. (Disc) (I) (GAGR) (Inductive)
Contrapositive: verse?

179. (If) ye be (not) converted and not become as little children, (then) ye shall not enter into the kingdom of heaven. (Disc) (I) (MAGR-F) (Inductive)
Contrapositive: verse?

180. (If one) shall humble himself as this little child, (then) the same is the greatest in the kingdom of heaven. (Disc) (I) (MAGR-N)
Contrapositive: verse???—see # 16, # 511, # 512, 224, 225 for different condition, but similar conclusion – The converse holds because of definition from "Whosoever therefore shall humble himself as this little child, the same is greatest in the kingdom of heaven."

181. (If one) shall receive one such little child in my name, (then that one) receiveth me. (Disc) (I) (MAGR-N&F) (1 person)
Contrapositive: verse? – See # 479 for same

182. (If one) shall (harm) one of these little ones which believe in me, (then) it were better for him that a millstone were hanged about his neck and that he were drowned in the depth of the sea. (Disc) (I) (MAMR) (1 person)
Contrapositive: verseF?—see # 883 for the same

183. (If one is a) man by whom the offence cometh, (then) woe unto that man. (Disc) (I) (MAGR-N&F)
Contrapositive: verse?—see # 882 for same

184. If thy hand or thy foot offend thee, (then) cut it off and cast them from thee. (Disc) (MAMR) (INFERRED—"IF" given)
Contrapositive: verseF?—#486 for "hand" and "cut" only

185. If thine eye offend thee, (then) pluck it out and cast it from thee. (Disc) (MAMR)
Contrapositive: verseF?

186. If a man have an hundred sheep and one of them be gone astray, (then) he doth leave the ninety and nine and goeth into the mountains and seeketh that which is gone astray. (Disc) (Analogy-literal)) (MAMR)
Contrapositive: verseF?

187. If so be that he (a man) find it (a sheep gone astray), (then) he rejoiceth more of that sheep, than of the ninety and nine which went not astray. (Disc) (Analogy) (MAMR- literally) (1 person)
Contrapositive: verseF?—see # 846

{
188. If thy brother shall trespass against thee, (then) go and tell him his fault between thee and him alone. (Disc) (MAMR) (1 person)
Contrapositive: verseF?—see # 884 for very similar, but not same conclusion as the other involves "rebuking"

189. If he (thy brother) shall hear thee, (then) thou hast gained thy brother. (Disc) (MAMR) (2 persons)
Contrapositive: verse?

190. If he (thy brother) shall not hear thee, then take with thee one or two more, then that in the mouth of two or three witnesses every word may be established. (Disc) (MAMR) (2 or more persons)
Contrapositive: verseF?

191. If he shall neglect to hear them (one or two more witnesses), (then) tell it unto the church. (Disc) (MAMR)
Contrapositive: verseF?

192. If he neglect to hear the church, (then) let him be unto thee as an heathen man and a publican. (Disc) (MAMR)
Contrapositive: verse?
}-grouping Conditional/Procedurally from #187–191, but not STINV, though negation of conclusion is used (if p, then q follwed by if ~q, then s, etc.)

193. #160 repeated (except that shalt is given as shall)

194. #161 repeated (except that shalt is given as shall)

195. If two of you (disciples) shall agree on earth as touching any thing that they shall ask, (then) it shall be done for them of my Father which is in heaven. (Disc) (MAGR-F)
Contrapositive: verse?

196. If two or three are gathered together in my name, (then) there am I in the midst of them. (Disc) (MAGR-N)
Contrapositive: verse?

197. If my brother shall sin against me, (then) I (should) forgive him seventy times seven. (Disc-Peter) (MAMR) (2 persons)

Contrapositive: verseF? (the contrapos. would seem to be apparently not true by human logic)—see # 522 for similar but adds praying as a condition

198. If ye from your hearts forgive not every one his brother their trespasses, then so likewise shall my heavenly Father do also unto you (where his lord was wroth, and delivered him to the tormentors, till he should pay all that was due unto him). (Disc-Peter) (MAGR-F)
Contrapositive: verse?—see # 46 for very similar

199. (If) a man leave father and mother, and shall cleave unto his wife, (then) they twain shall be one flesh. (Relig/Mult.) (I) (MAGR-N)
Contrapositive: verse?—see # 495 for same

200. (If) God hath joined (man and wife) together, (then) let not man put (them) asunder. (Relig/Mult.) (I) (GAMR)
Contrapositive: verseF?—see # 496 for same

201. (If one) shall put away his wife, except it be for fornication, and shall marry another, (then that one) committeth adultery. (Relig./Mult) (I) (MAMR)
Contrapositive: verseF? – See # 497 and # 498 for similar (those don't have fornication) and # 31 for similar and # 874 for same

202. (If one) marrieth her which is put away, (then that one) doth commit adultery. (Relig/Mult.) (I) (MAMR)
Contrapositive: verseF?—see # 32 for same

203. (If a man is one) to whom it is given, (then that man) can receive this saying (on marriage). (Disc) (I) (GAMR)
Contrapositive: verse?

204. (If a man) is able to receive it, (then) let him receive it. (Disc) (I) (GAGR and MAGR—see preceding verse, Matt 19:12a)—(2 persons)
Contrapositive: verseF?

205. If thou wilt enter into life, (then) keep the commandments. (Rich young ruler) (MAMR) (1 person)
Contrapositive: verse?—see # 94 and #471 for similar condition and similar conclusion (See # 1147 for close to converse)

206. If thou wilt be perfect, (then) go and sell that thou hast, and give to the poor. (Rich young ruler) (MAMR) (1 person)
Contrapositive: verse?

207. If (thou) go and sell that thou hast, and give to the poor, (then) thou shalt have treasure in heaven. (Rich young ruler) (MAMR) (1 person) (Inductive)
Contrapositive: verse? (Would seem to be true as contrapositive for rich young ruler, but not in general)—see # 501 and # 910 for same and see # 785 for similar

208. If thou wilt be perfect, (then) go and sell that thou hast, and give to the poor, . . . and come and follow me.(Rich young ruler) (MAMR) (1 person)
Contrapositive: verse? (Would seem to be true as contrapositive for rich young ruler, but not in general)

209. (If a man is rich), (then) that man shall hardly enter into the kingdom of heaven. (Disc) (I) (MAGR-F) (Inductive)—see # 503 for similar and # 911 for same
Contrapositive: verseF?

210. (If) ye . . . have followed me, (then) in the regeneration when the Son of man shall sit in the throne of his glory, ye also shall sit upon twelve thrones, judging the twelve tribes of Israel. (Disc) (I) (MAGR-F)
Contrapositive: verse?—see # 991 for very similar

211. (If) one hath forsaken houses, brethren, or sisters, or father, or mother, or wife, or children, or lands, for my name's sake, (then that one) shall receive an hundredfold, and shall inherit everlasting life. (Disc) (I) (MAGR-F)
Contrapositive: verseF?—see # 94 and # 471 for conclusion as condition

212. When he had agreed with the labourers for a penny a day, (then) he sent them into his vineyard. (Disc) (Analogy-literal) (MAMR- 1 person)
Contrapositive: verseF?

213. (If) ye go also into the vineyard, (then) whatsoever is right I will give you. (Disc) (I) (Analogy-literal) (MAMR- 2 persons) (Inductive)
Contrapositive: verse?

214. (If) ye go also into the vineyard, (then) whatsoever is right, that shall ye receive. (Disc) (I) (Analogy-literal) (MAMR- 2 persons) (Inductive)
Contrapositive: verse?

215. When even was come, (then) the lord of the vineyard saith unto his steward, Call the labourers, and give them their hire, beginning from the last unto the first. (Disc) (Analogy-literal) ((None)
Contrapositive: verseF?

216. When they came that were hired about the eleventh hour, they received every man a penny. (Disc) (Analogy-literal) (MAMR – 1 group and another person)
Contrapositive: verseF?

217. When the first came, they supposed that they should have received more. (Disc) (Analogy-literal) (MAMR- 1 group)
Contrapositive: verseF?

218. When the first came and they supposed that they should have received more, (then) they likewise received every man a penny. (Disc) (Analogy-literal) ((MAMR – 1 group and another person)
Contrapositive: verseF?

219. When they (the first hired) received it (a penny), they murmured against the goodman of the house. (Disc) (Analogy-literal) (GAMR)
Contrapositive: verseF?

220. (If) thou didst agree with me for penny, (then) I do thee no wrong, Friend. (Disc) (Analogy-literal) (MAGR-N)
Contrapositive: verseF?

221. (If) one is the last, (then) one shall be first. (Disc) (Analogy-literal) (MAGR-F) (Inductive)
Contrapositive: verseF?—see Luke 13:30 (not in this list) in that the qualification is made that "there are last which shall be first and there are first which shall be last" (i.e. – not all)

222. (If) one is the first, (then) one shall be last. (Disc) (Analogy-literal) (MAGR-F) (Inductive)
Contrapositive: verseF? – See # 478 for very similar see Luke 13:30 (not in this list) in that the qualification is made that "there are last which shall be first and there are first which shall be last" (i.e. – not all)

223. (When) we go up to Jerusalem, (then) the Son of man shall be betrayed unto the chief priests and unto the scribes and they shall condemn him to

death and shall deliver him to the Gentiles to mock, and to scourge, and to crucify him. (Disc) (MAMR, though Christ is involved in the going up)
Contrapositive: verseF?—see # 507 for similar

224. (If you) let (one) . . . be your minister, (then that one) will be great among you. (Disc) (MAMR) (Inductive)
Contrapositive: verseF??? (definition of great is?)—see 16, 180, 511, 512, 225

225. (If you) let (one) . . . be your servant, (then that one) will be chief among you. (Disc) (MAMR) (Inductive)
Contrapositive: verseF? (NOTE how "Conditional/Procedural" this is in human reasoning commonly, though not by strict logic depending on definition of chief)—see 224, 16, 180, 511, 512 (see # 990 for same)

226. (When ye) go into the village over against you, (then) straightway ye shall find an ass tied, and a colt with her. (Disc) (MAMR?)
Contrapositive: verseF?—see #515

227. (When) straightway ye shall find an ass tied, and a colt with her, (then) loose them and bring them unto me. (Disc) (MAMR?)
Contrapositive: verseF?—see # 516

228. If any man say aught unto you, (then) ye shall say, The Lord hath need of them. (Disc) (MAMR)—Contrapositive: verseF?—see # 517

229. If any man say aught unto you, (and) ye shall say, The Lord hath need of them, (then) straightway he will send them. (Disc) (MAMR)
Contrapositive: verseF?

230. (When) thy King cometh unto thee, (He shall be) meek, and sitting upon an ass, and a colt the foal of an ass. (Disc) (GAGR)
Contrapositive: verse?

231. If ye have 'faith', and doubt not, (then) ye shall not only do this which is done to the fig tree. (Disc) (MAGR-N&F) (Inductive)
Contrapositive: verse?

232. If ye have 'faith', and doubt not (and if) ye shall say unto this mountain, Be thou removed, and be cast into the sea, (then) it shall be done. (Disc) (Analogy- literal) (MAGR-N&F) – See # 231 above since the condition of # 231 is also implied as a condition in #232. – See # 520
Contrapositive: verse?

233. (If) shall ask in prayer (any) thing (and) believe, (then) ye shall receive. (Disc) (MAGR-N&F)
Contrapositive: verseF?—see #521 for very similar and # 60 and Luke 11:10 for somewhat similar

234. If ye (chief priests and elders of the temple) tell me one thing (that) I will ask you (was the baptism of John from heaven or of men), (then) I in likewise will tell you by what authority I do these things. (Relig) (MAGR-N) (Inductive)
Contrapositive: verseF?—see also sort of INVERSE in Matt 21: 27, but requires "since" not "if"—see # 525 for highly similar

235. (If) one is a publican or a harlot, (then that one) go(es) into the kingdom of God before you. (Relig) (MAGR-F)
Contrapositive: verse?

236. (When) John came unto you in the way of righteousness, ye believed him not. (Relig) (MAMR)
Contrapositive: verseF?

237. (When) John came unto you in the way of righteousness, the publicans and harlots believed him. (Relig) (MAMR)
Contrapositive: verseF?

238. (When) ye had seen it (John coming unto you in the way of righteousness), ye repented not afterward, that ye might believe him. (Relig) (MAMR)
Contrapositive: verseF?

239. . . . when the time of the fruit drew near, he sent his servants to the husbandmen, that they might receive the fruits of it. (Relig) (Analogy-literal) (GAGR)
Contrapositive: verseF?—#527 essentially the same

239a) Matt 21: 35—very similar to #528, but not exact same wording – (when) . . . the husbandmen took his servants, (then he) . . . beat one (of his servants), and killed another, and stoned another.

240. (When) he sent other servants more than the first, . . . they did unto them likewise. (Relig) (Analogy-literal) (GAMR)
Contrapositive: verseF?—see #529 for similar, but not exactly

241. . . . when the husbandmen saw the son, they said among themselves, this is the heir, come, let us kill him, and let us seize on his inheritance. (Relig) (Analogy-literal) (MAMR)
Contrapositive: verseF?

242. (If one) shall fall on this stone (the one "which the builders rejected"), (then that one) shall be broken. (Relig) (Analogy-symbolic) (MAGR)
Contrapositive: verseF? – See # 941 for same

243. (If the stone) shall fall on (one), (then) the stone will grind (that one) to powder. (Relig) (Analogy-symbolic) (GAGR) (Inductive)
Contrapositive: verse? – See # 942 for same

244. (When the king) sent forth his servants to call them that were bidden to the wedding, (then) they (that were bidden) would not come. (Relig) (Analogy-literal) (MAMR or if symbolic GAMR)
Contrapositive: verseF?

245. (When the king) sent forth other servants, . . . they (those who were bidden) made light of it, and went their ways, one to his farm, another to his merchandise: and the remnant took his servants, and entreated them spitefully, and slew them. (Relig) (Analogy-literal) (MAMR or if symbolic GAMR)
Contrapositive: verseF?

246. When the king heard thereof, (then) he was wroth and he sent forth his armies, and destroyed those murderers, and burned up their city. (Relig) (Analogy-literal) (MAMR, if symbolic GAGR)
Contrapositive: verseF?

247. When the king heard thereof, he was wroth and he sent forth his armies, and destroyed those murderers, and burned up their city, Then saith he (the king) to his servants, The wedding is ready, but they which were bidden were not worthy.
Contrapositive: verseF? (A use of A−>B−>C****) (Relig) (Analogy-literal) (MAMR, if symbolic GAGR)

248. (Since) The wedding is ready, but they which were bidden were not worthy, go ye therefore into the highways, and as many as ye shall find, bid to the marriage. (Relig) (Analogy-literal) (MAMR—both literal and symbolic)
Contrapositive: verseF?

249. (When) those servants went out into the highways, and gathered together all as many as they found, both good and bad, (then) the wedding was furnished with guests. (Relig) (Analogy-literal) (MAMR, if symbolic MAGR)
Contrapositive: verse?

250. When the king came in to see the guests, (then) he saw there a man which had not a wedding garment. (Relig) (Analogy-literal) (MAMR, if symbolic GAGR)
Contrapositive: verse?

251. When the king came in to see the guests and he saw there a man which had not a wedding garment, (then) he saith unto him, Friend, how camest thou in hither not having a wedding garment. (Relig) (Analogy-literal) (MAMR, if symbolic GAGR) ? (A use of A->B->C****)
Contrapositive: verse?

252. When the king came in to see the guests and he saw there a man which had not a wedding garment, (and) he saith unto him, Friend, how camest thou in hither not having a wedding garment, Then said the king to the servants, Bind him hand and foot, and take him away, and cast him into outer darkness; there shall be weeping and gnashing of teeth. (Relig) (Analogy-literal) (MAMR, if symbolic GAGR)? (A use of A->B->C->D***)
Contrapositive: verse?

253. (Since the image and superscription on the penny) is Caesar's, Render therefore unto Caesar the things that are Caesar's; and unto God the things that God's. (Relig) (MAMR)
Contrapositive: verseF?—see # 943 for same

254. (If the scribes and the Pharisees) therefore bid you do whatsoever, (then) that observe and do. (Mult) (Disc) (MAMR) (Inferred)
Contrapositive: verseF?

255. (Since) they (the scribes and the Pharisees) say, and do not, (then) do not ye after their works. (Mult) (Disc) (MAMR) (Inferred)
Contrapositive: verseF?

256. (When) they (the scribes and the Pharisees) do all their works, (then) they do (their works) for to be seen of men. (Mult) (Disc) (MAMR) (Inferred)
Contrapositive: verse?

257. (Since) one is your Master, even Christ, (then) be not ye called Rabbi. (Mult) (Disc) (GAMR) (Inferred)
Contrapositive: verse?

258. (Since) one is your Father, which is in heaven, (then) call no man your father upon the earth. (Mult) (Disc) (GAMR) (Inferred)
Contrapositive: verseF?

259. (Since) one is your Master, even Christ, (then) neither be not ye called masters. (Mult) (Disc) (GAMR) (Inferred)
Contrapositive: verse?

260. (If one) shall be your servant, (then that one) is the greatest among you. (Mult) (Disc) (MAMR) (Inferred)
Contrapositive: verse?—see 16, 180, 224, 225, 511, 512 (see # 989 for condition as younger" not "servant", but same conclusion)

{
261. (If one) shall exalt himself, (then that one) shall be abased.(Mult) (Disc) (MAGR) (Inferred)
Contrapositive: verse?—see # 827 for same

262. (If one) shall humble himself, (then that one) shall be exalted.(Mult) (Disc) (MAGR) (Inferred)
Contrapositive: verse?—see # 828 for same } STINV close, but not exalted is not the same as humble given middle ground of neutral.

263. (Since) ye (scribes and Pharisees, hypocrites!) shut up the kingdom of heaven against men, (then) woe unto you. (Mult) (Disc) (MAGR) (Inferred)
Contrapositive: verseF?

264. (Since) ye (scribes and Pharisees, hypocrites!) neither go in yourselves, neither suffer ye them that are entering to go in, (then) woe unto you. (Mult) (Disc) (MAGR) (Inferred)
Contrapositive: verseF?

265. (Since) ye (scribes and Pharisees, hypocrites!) devour widows' houses and make long prayer, (then) woe unto you. (Mult) (Disc) (MAGR) (Inferred)
Contrapositive: verseF? A—> B—see # 538 for very similar (not exact on condition or conclusion)

266. (Since) ye (scribes and Pharisees, hypocrites!) devour widows' houses and make long prayer, therefore ye shall receive the greater damnation. (Mult) (Disc) (MAGR) (Inductive)
Contrapositive: verseF? (Use of same condition in # 265 and 266 with different conclusion) A−>C—see # 538 for essentially the same (though condition is slightly different)

267. (Since) ye (scribes and Pharisees, hypocrites!) compass sea and land to make one proselyte and when he is made, ye make him twofold more the child of hell than yourselves, (then) woe unto you. (Mult) (Disc) (MAGR) (Inferred)
Contrapositive: verseF?

268. . . . when he (one is being proselytized by scribes and Pharisees) is made, ye make him twofold more the child of hell than yourselves. (Mult) (Disc) (MAMR) –see #267
Contrapositive: verse?

269. (Since) ye (blind guides) say, "Whosoever shall swear by the temple, it is nothing, but whosoever shall swear by the gold of the temple, he is a debtor, (then) woe unto you. (Mult) (Disc) (MAGR) (Inferred)
Contrapositive: verseF?

270. (If one) shall swear by the altar, (then) it is nothing. (Mult) (Disc) (MAGR)
Contrapositive: verseF?

271. (If one) sweareth by the gift that is upon it (the altar), (then) he is guilty. (Mult) (Disc) (MAGR)
Contrapositive: verseF?

272. (Since) the altar that sanctifieth the gift is greater than the gift, therefore whoso shall swear by the altar, sweareth by it, and by all things thereon. (Mult) (Disc) (None)(INFERRED- CONDITION IS INFERRED FROM APPARENT RHETORICAL COMMENT PRECEDING THE "THEREFORE")
Contrapositive: verse?

273. (If one) shall swear by the temple, (then that one) sweareth by it (the temple) and by him that dwelleth therein. (Mult) (Disc) (Inferred) (MAMR)
Contrapositive: verseF?

274. (If one) shall swear by heaven, (then that one) sweareth by the throne of God and by him that sitteth thereon. (Mult) (Disc) (Inferred) (MAMR)
Contrapositive: verseF?

275. (Since) ye (scribes and Pharisees, hypocrites!) pay tithe of mint and anise and cumin, and have omitted the weightier matters of the law, judgment, mercy, and 'faith', (then) woe unto you. (Mult) (Disc) (MAGR) (Inferred)
Contrapositive: verseF?—see # 756 for essentially the same (slightly different wording)

276. (Since) ye (scribes and Pharisees, hypocrites!) make clean the outside of the cup and of the platter, but within they are full of extortion and excess, (then) woe unto you. (Mult) (Disc) (MAGR) (Inferred)
Contrapositive: verseF?

277. (If a) blind Pharisee, cleanse(s) first that which is within the cup and platter, (then) the outside of them may be clean also. (Multi) (Disc) (MAGR) (Inferred) (Analogy- literal)
Contrapositive: verseF? (literal and symbolic)

278. (Since) ye (scribes and Pharisees, hypocrites!) are like unto whited sepulchres, which indeed appear beautiful outward, but are within full of dead men's bones, and of all uncleanness, (then) woe unto you. (Mult) (Disc) (MAGR) (Inferred)
Contrapositive: verseF?

279. Because ye (scribes and Pharisees, hypocrites!) build the tombs of the prophets, and garnish the sepulchres of the righteous and say, If we had been in the days of our fathers, we would not have been partakers with them in the blood of the prophets, (then) woe unto you. (Mult) (Disc) (MAGR) (Inferred)
Contrapositive: verseF?—see # 761 for the first half of the condition ("build") and the same conclusion- FORM: A and B implies C whereas # 761 is A and D implies C.

280. (Since) ye build the tombs of the prophets, and garnish the sepulchres of the righteous and say, If we had been in the days of our fathers, we would not have been partakers with them in the blood of the prophets, wherefore ye be witnesses unto yourselves, that ye are the children of them which killed the prophets. (Mult) (Disc) (MAMR) (Need to check IF THIS IS INDEED LOGIC or just LOOSELY CONNECTED)
Contrapositive: verse?

281. (Since) ye be witnesses unto yourselves, that ye are the children of them which killed the prophets, (then) fill ye up the measure of your fathers. (Mult) (Disc) (MAMR)
Contrapositive: verseF?

282. (Since) ye (are) serpents, ye generation of vipers, wherefore, behold I send unto you prophets, and wise men, and scribes . . . (Mult) (Disc) (MAGR) (INFERRED)
Contrapositive: verseF?

283. (Since), behold I send unto you prophets, and wise men, and scribes: (then) . . . some of them ye shall kill and crucify and some of them shall ye scourge in your synagogues, and persecute them from city to city. (Multi) (Disc) (GAMR) (INFERRED)
Contrapositive: verseF?

284. (Since) some of them (prophets, and wise men, and scribes), ye shall kill and crucify and some of them shall ye scourge in your synagogues, and persecute them from city to city, (then) upon you may come all the righteous blood shed upon the earth, from the blood of righteous Abel unto the blood of son of Barachias, whom ye slew between the temple and the altar. (Multi) (Disc) (MAGR) (INFERRED)—see # 98 and # 763 for similar and 613 for "inverse like"
Contrapositive: verse?

285. (If one) cometh in the name of the Lord, (then that one) is blessed. (Multi) (Disc) (MAGR) (INFERRED)
Contrapositive: verse?—see # 821 for same

286. (Since) many shall come in my name, saying, I am Christ, and deceive many, (then you) take heed that no man deceive you. (MAMR) (Disc) (INFERRED)
Contrapositive: verseF?—see # 540

287. (When) ye hear of wars and rumours of wars, (then) see that ye be not troubled. (MAMR) (Disc) (INFERRED)
Contrapositive: verseF?

288. (Since) all these things must come to pass, (then) see that ye be not troubled. (GAMR) (Disc) (INFERRED)
Contrapositive: verseF?

289. (Since) nation shall rise against nation, and kingdom against kingdom: and there shall be famines, pestilences, and earthquakes in divers places, (then) the end is not yet. (1st part is MAGR, but 2nd part is GAGR) (Disc) (INFERRED)
Contrapositive: verse?—see # 543 for different condition, but same conclusion

290. (When) nation shall rise against nation, and kingdom against kingdom: and there shall be famines, pestilences, and earthquakes in divers places, then shall they deliver you up to be afflicted, and shall kill you and ye shall be hated of all nations for my name's sake. (1st part is MAMR, but 2nd part is GAMR) (Disc)
Contrapositive: verseF?—see # 551 for different condition, but also has ending of hated . . . for my name's sake.

291. (When) nation shall rise against nation, and kingdom against kingdom: and there shall be famines, pestilences, and earthquakes in divers place, then shall many be offended, and shall betray one another, and shall hate one another. (Disc) (1st part is MAMR, but 2nd part is GAMR) (A -> B in #290, and A -> C in #291)
Contrapositive: verseF?

292. (When) nation shall rise against nation, and kingdom against kingdom: and there shall be famines, pestilences, and earthquakes in divers place, (then) many false prophets shall rise, and shall deceive many. (Disc) (1st part is MAMR, but 2nd part is GAMR) (A -> B in #290, and A -> C in #291 and A-> D in #292) (INFERRED)
Contrapositive: verseF?

293. Because iniquity shall abound, (then) the love of many shall wax cold. (Disc) (MAMR)
Contrapositive: verse?

294. (If) one shall endure to the end, (then that one) shall be saved. (Disc) (MAGR) (INFERRED)
Contrapositive: verse?—see # 552 for same and see # 587 for same conclusion, but different condition—see # 84d) for same as well

295. (When) this gospel of the kingdom shall be preached in all the world for a witness unto all nations, (then) shall the end come. (Disc) (GAGR, but also MAGR) (INFERRED)
Contrapositive: verse?

296. When ye therefore shall see the abomination of desolation, spoken of by Daniel the prophet, (then) stand in the holy place, (whoso readeth, let him understand). (Disc) (GAMR) (INFERRED) (A -> B, see below)
Contrapositive: verseF?—see # 553

297. When ye therefore shall see the abomination of desolation, spoken of by Daniel the prophet, then let them which be in Judaea flee into the mountains. (Disc) (GAMR) (INFERRED) (A-> C, see #296)
Contrapositive: verseF?—see # 553

298. When ye therefore shall see the abomination of desolation, spoken of by Daniel the prophet, (then) let him which is on the housetop not come down to take any thing out of his house neither let him which is in the field return back to take his clothes. (Disc) (GAMR) (INFERRED) (A-> D, see #296)
Contrapositive: verseF?—see # 553

299. (If one is) with child (at the "abomination of desolation"), (then) woe unto (that one). (Disc) (MAGR-F) (INFERRED)
Contrapositive: verseF?—see # 554

300. (If one is) giving suck in those days, (then) woe unto (that one). (Disc) (MAGR-F) (INFERRED)
Contrapositive: verseF?—see # 554

301. (Since) then shall be great tribulation, such as was not since the beginning of the world to this time, no nor ever shall be, (then) pray ye that your flight be not in the winter, neither on the sabbath day. (Disc) (GAMR) (INFERRED)
Contrapositive: verseF?

301 a—see # 555

302. (If) those days should (not) be shortened, (then) there should no flesh be saved. (Disc) (GAGR) (INFERRED)
Contrapositive: verseF?

303. (When) for the elect's sake those days shall be shortened, then if any man shall say unto you, Lo, here is Christ, or there: (then you) believe it not. (Disc) (GAMR & MAMR) (INFERRED) (A and B -> C) or possibly A then B then C
Contrapositive: verseF?

304. (Since) there shall arise false Christs, and false prophets, and shall shew great signs and wonders; in somuch that, if it were possible, they shall deceive the very elect, (then) believe it (that Christ is here or there) not. (Disc) (MAMR) (INFERRED) (see # 303 above—D -> C from above)—see # 286 and # 504 for similar
Contrapositive: verseF?—see # 1045 to show that one can see signs and wonders, even if not believe.

305. (Since) I have told you before, (then) if they shall say unto you, Behold, he is in the desert, go not forth: (or) behold, he is in the secret chambers, believe it not. (Disc) (GAMR and MAMR)) (INFERRED) (E-> (F->G or H))
Contrapositive: see # 303 above overall (E-> (F->G or H)) for (GAMR) as verseF? and (E->G or H) for (MAMR) as verseF?

306. (Since) as the lightning cometh out of the east, and shineth even unto the west; so shall also the coming of the Son of man be, (then) if they shall say unto you, Behold, he is in the desert, go not forth: (or) behold, he is in the secret chambers, believe it not. (Disc) (GAMR) (INFERRED) (see # 303 above (I -> (F->G or H))
Contrapositive: (I -> (F->G or H)) verseF? and (F->G or H) see #305 as it is the same)—see # 891 for essentially same as # 306, except that the conclusion in # 891 does not have the "secret chamber" portion.

307. (Since) wheresoever the carcase is, there will the eagles be gathered together, (then) so shall also the coming of the Son of man be. (Disc) (GA__) (INFERRED) ((B->C)-> A)
Contrapositive: verse?

308. (If) the carcase is (wheresoever), there will the eagles be gathered together. (Disc) (____) (INFERRED) (B ->C)
Contrapositive: verseF?—see # 1494 for converse by definition

309. (When) the tribulation of those days (occurs), (then) shall the sun be darkened, and the moon shall not give her light, and the stars shall fall from heaven, and the powers of the heavens shall be shaken. (Disc) (GAGR) (INFERRED) (A-> B and C and D and E)
Contrapositive: verse?

310. (When) . . . the sun (shall) be darkened, and the moon shall not give her light, and the stars shall fall from heaven, and the powers of the heavens shall

be shaken, then shall appear the sign of the Son of man in heaven. (Disc) (GAGR) (B &C & D & E -> F)—see #309
Contrapositive: verse?

311. (When) the sign of the Son of man in heaven (shall appear), then shall all the tribes of the earth mourn, and they shall see the Son of man coming in the clouds of heaven with power and great glory. (Disc) (GAMR) (F -> G and H)—see #310
Contrapositive: verse?

312. (When) they (the tribes of the earth) shall see the Son of man coming in the clouds of heaven with power and great glory, (then) he shall send his angels with a great sound of a trumpet. (Disc) (MAGR) (INFERRED) (H -> I)—see # 311
Contrapositive: verseF? (see #311)

313. (When) he shall send his angels with a great sound of a trumpet, (then) they (the angels) shall gather together his elect from the four winds, from one end of heaven to the other. (Disc) (GAGR) (INFERRED) (I -> J)—see #312
Contrapositive: verse?

314. (When) the branch is yet tender, and putteth forth leaves, (then) ye know that summer is nigh. (Disc) (GAMR) (INFERRED)
Contrapositive: verseF?—branches could be tender without human knowledge—see # 973 for essentially the same though condition is stated with different words that have same effect

315. When ye shall see all these things, (then) know that it is near, even at the doors. (Disc) (MAGR) (INFERRED)
Contrapositive: verse?

316. (When) shall also the coming of the Son of man be, then shall two be in the field; the one shall be taken, and the other left. (Disc) (GAGR) (INFERRED) A->B
Contrapositive: verse?

317. (When) shall also the coming of the Son of man be, then shall two women shall be grinding at the mill; the one shall be taken, and the other left. (Disc) (GAGR) (INFERRED) A->C—see #316
Contrapositive: verse?

318. Since shall also the coming of the Son of man be, (and) then shall two women shall be grinding at the mill; the one shall be taken, and the other left, watch therefore. (Disc) (GAMR) (C−> D)—see # 317
Contrapositive: verseF? (this would seem to be False potentially as one could not watch, but since no man knows how Christ is returning, then the contrapositive's conclusion would seem to always be false)

319. (Since) ye know not what hour your Lord doth come, watch therefore. (Disc) (E −> D)—see #318 (Two slightly different conditions implying the same conclusion, though the main difference in the conditions is that of God's action and man's knowledge of that action) (MAMR)
Contrapositive: verseF? **** (This would seem to be False potentially as one could not watch, but since no man knows when Christ is returning, then the contrapositive's conclusion would seem to always be false)—see # 567 for similar thought, though different words in conclusion and condition

320. If the goodman of the house had known in what watch the thief would come, (then) he would have watched, and would not have suffered his house to be broken up. (Disc) (Analogy-literal and symbolic) (MAMR) (INFERRED, though IF is given) (A−>B)
Contrapositive: verseF?

321. (Since) if the goodman of the house had known in what watch the thief would come, (then) he would have watched, and would not have suffered his house to be broken up, therefore be ye ready. (Disc) (MAMR) (INFERRED, though therefore is given) (Analogy-literal) (A−>B) −> C—see # 320
Contrapositive: verse? (seems to be true by definition taking "and" literally)

322. (Since) in such an hour as ye think not the Son of man cometh, therefore be ye also ready. (Disc) (GAMR) (INFERRED) (D−>C) (#321 and #320)
Contrapositive: verseF? (Seems to be false as Christ's Return is not dependent on our readiness

323. (If) in such an hour as ye think not, (then) the Son of man cometh. (Disc) (MAGR-F) (INFERRED) (D = E−>F) (see #323)
Contrapositive: verseF? (seems to be false as man can think that Christ is returning but He did not as history has shown)

324. (If one's) . . . lord when he cometh shall find (that one) so doing (being 'faith'ful and wise), (then) blessed is that servant (one). (Disc) (GAGR) (Analogy-literaliteralymbolic) (INFERRED)
Contrapositive: verseF?

325. When he (a servant's lord) cometh, (then) he (a servant's lord) shall find him so doing (being 'faith'ful and wise). (Disc) (GAGR) (Analogy- literal and symbolic) (INFERRED)
Contrapositive: verseF?

326. (If one's) . . . lord when he cometh shall find (that one) so doing (being 'faith'ful and wise), (then that one's lord) shall make him ruler over all his goods. (Disc) (MAMR - GAGR) (Analogy – literaliteralymbolic) (INFERRED)
Contrapositive: verse?

327. If that evil servant shall say in his heart, My Lord delayeth his coming; and shall begin to eat and drink with the drunker; (then) the lord of that servant shall come in a day when he looketh not for him, and in hour that he is not aware of, and shall cut him asunder, and appoint him his portion with the hypocrites . . . (Disc) (MAMR - MAGR) (Analogy- literaliteralymbolic) (INFERRED)
Contrapositive: verse? (given the when he looketh not for him would seem to make this true always.

328. (W)hen he (that evil servant) looketh not for him (his lord), (then) the lord of that servant shall come in a day, . . . and in hour that he is not aware of, and shall cut him asunder, and appoint him his portion with the hypocrites. (Disc) (MAMR - MAGR) (Analogy- literal& symbolic) (INFERRED)
Contrapositive: verse? (seems to be true by: aware−> looking for)

329. (When) the lord (of that evil servant) . . . shall cut him asunder, and appoint him his portion with the hypocrites, (then) there shall be weeping and gnashing of teeth. (Disc) (MAMR – GAMR) (Analogy- literaliteralymbolic) (INFERRED)
Contrapositive: verse?

{
330. (If they were) foolish (virgins), (then they) took their lamps, and took no oil with them. (Disc) (MAMR- MA?) (Analogy-literaliteralymbolic) (INFERRED)

Contrapositive: verse? (If OR holds to inclusive "OR" then the contrapositive seems true)

331. (if they were) wise virgins, (then they) took oil in their vessels with their lamps. (Disc) (MAMR- MA?) (Analogy – literaliteralymbolic) (INFERRED)
Contrapositive: verse? STINV- inverse
}

332. (When) the bridegroom tarried, (then) they all slumbered and slept. (Disc) (MAMR – GAMR) (Analogy – literaliteralymbolic) (INFERRED)
Contrapositive: verseF?

333. (When) midnight came, (then) there was a cry made . . . (Disc) (GAMR – GAGR) (Analogy- literaliteralymbolic) (INFERRED)
Contrapositive: verseF?

334. . . . (Since) Behold the bridegroom cometh, (then) go ye out to meet him. (Disc) (MAMR- GAMR) (Analogy- literaliteralymbolic) (INFERRED)
Contrapositive: verseF? (God's events seem not to be dependent on man's actions.)

335. . . . (When) at midnight there was a cry made, Behold the bridegroom cometh, go ye out to meet him, Then all those virgins arose, and trimmed their lamps. (Disc) (MAMR- GAMR) (Analogy- literaliteralymbolic) (INFERRED)
Contrapositive: verseF?

336. (Since) our lamps are gone out, (then) Give us (the foolish) of your (the wise) oil. (Disc) (MAMR – MAMR) (Analogy – literaliteralymbolic) (INFERRED)
Contrapositive: verseF?

337. (When) the bridegroom came; (then) they that were ready went in with him to the marriage and the door was shut. (Disc) (MAMR – GAGR) (Analogy – literal and symbolic) (INFERRED)
Contrapositive: verse? (seems true by "default" of terms used equivalently)

338. (When) afterward came also the other virgins, saying, Lord, Lord, open to us, (then) but he answered and said, Verily I say unto you, I know you not. (Disc) (MAMR – MAGR) (Analogy – lit & symb) (INFERRED) (See #816 for similar)
Contrapositive: verse?

339. (Since) ye know neither the day nor the hour wherein the Son of man cometh, watch therefore. (Disc) (MAMR) (INFERRED)
Contrapositive: verse?

340. (If) one is a servant of a man, (then) one receives talents according to his several ability. (Disc) (Analogy – Literaliteralymbolic) (MAMR – MAGR) (INFERRED)
Contrapositive: verse?

341. (When) he (the man) gave unto one five talents, to another two, and to another one; to every man according to his ability; and straightway (the man) took his journey. (Disc) (Analogy – literal) (MAMR – GAGR) (INFERRED)
Contrapositive: verseF?

342. (When) straightway (the man) took his journey, Then he that had received the five talents went and traded with the same, and made them other five talents. (Disc) (Analogy- literal) (MAMR – GAMR) (INFERRED) A−>B
Contrapositive: verseF?

343. (When) straightway (the man) took his journey, (then) and likewise he that had received two, he also gained other two. (Disc) (Analogy – literal) (MAMR – GAMR) (INFERRED) A−>C
Contrapositive: verseF?

344. (When) after a long time (passed), (then) the lord of those servants cometh, and reckoneth with them. (Disc) (Analogy- literal) (GAMR – GAGR) (INFERRED)
Contrapositive: verseF?

345. (Since) thou (the servant with five talents and also the servant with two talents) hast been 'faith'ful over a few things, (then) I (the man) will make thee ruler over many things. (Disc) (MAMR – MAGR) (Analogy – literal) (INFERRED) A−>B
Contrapositive: verseF?

346. (Since) the man (with one talent) knew that thou art an hard man, reaping where thou hast not sown, and gathering where thou hast not strawed, (then) I was afraid, and went and hid thy talent in the earth . . . (Disc) (MAMR – MAMR) (Analogy – literal) (INFERRED) C−>D
Contrapositive: verseF?

347. (Since) thou wicked and slothful servant, thou knewest that I (the lord of the servant) reap where I sowed not, and gather where I have not strawed, (then) thou oughtest therefore to have put my money to the exchangers . . . (Disc) (Analogy – literal) (MAMR- MAMR) (INFERRED) E−>F
Contrapositive: verseF?

348. . . . (if) thou oughtest therefore to have put my money to the exchangers, then at my coming I should have received mine own with usury. (Disc) (Analogy – literal) (MAMR- MAGR) (INFERRED) F−>G (Uses chaining of E−>F−>G))
Contrapositive: verse?

349. (Since the servant) was afraid, and went and hid thy (the lord's) talent in the earth: take therefore the talent from him, and give it unto him which hath ten talents. (Disc) (Analogy-literal) (MAMR – MAGR) (INFERRED) D−>H (Seems to uses chaining of C−>D−>H) (see # 352)
Contrapositive: verse?

{
350. (If) one hath, (then) unto (that one) shall be given (more) and he shall have abundance: . . . . (Disc) (Analogy-literal) (MA/GA-MR – MA/GA-GR) (INFERRED) (Inductive)
Contrapositive: verseF?—see # 125 for essentially the same

351. (If) one hath not, (then) unto (that one) shall be taken away even that he hath. (Disc) (Analogy-literal) (MAMR –MAGR) (INFERRED) (Inductive)—see # 126 and # 57
Contrapositive: verse?
} Close to INVERSE: STINV

352. (Since) unto every one that hath shall be given, therefore take the talent from him (the one with one talent), and give it unto him which hath ten talents. (Disc) (Analogy – literal) (MAMR – GAGR) (INFERRED) I−>H (see #349
Contrapositive: verseF? (by definition seems false)

353. When the Son of man shall come in his glory, and all the holy angels with him, then shall he sit upon the throne of his glory and before him shall be gathered all nations: and he shall separate them one from another, as a shepherd divideth his sheep from the goats: and he shall set the sheep on his right hand, but the goats on the left. (Disc) (GAGR)

Contrapositive: verse? (seems true by default since the negation of conclusion is always false)

354. (When) shall he sit upon the throne of his glory . . . . goats on the left, Then shall the King say unto them on his right hand, Come, ye blessed of my Father, inherit the kingdom prepared for you from the foundation of the world: (Disc) (GAGR)
Contrapositive: verse? (seems true by default since the negation of conclusion is always false)—see # 361 for same conclusion

355. (When) I was an hungred, (then) ye gave me meat . . . (Disc) (GAMR or MAMR depending upon "I" in context of passage as Christ explains later)
Contrapositive: verseF?

356. (When) I was a thirsty, (then) ye gave me drink . . . (Disc) (GAMR or MAMR depending upon "I" in context of passage as Christ explains later) (INFERRED)
Contrapositive: verseF?

357. (When) I was a stranger, (then) ye took me in . . . (Disc) (GAMR or MAMR depending upon "I" in context of passage as Christ explains later) (INFERRED)
Contrapositive: verseF?

358. (When) I was naked, (then) ye clothed me . . . (Disc) (GAMR or MAMR depending upon "I" in context of passage as Christ explains later) (INFERRED)
Contrapositive: verseF?

359. (When) I was a sick, (then) ye visited me . . . (Disc) (GAMR or MAMR depending upon "I" in context of passage as Christ explains later) (INFERRED)
Contrapositive: verseF?

360. (When) I was in prison, (then) ye came unto me . . . (Disc) (GAMR or MAMR depending upon "I" in context of passage as Christ explains later) (INFERRED)
Contrapositive: verseF?

361. Since ye gave me meat (when) I was hungred . . . ., Then shall the King say unto them on his right hand, Come, ye blessed of my Father, inherit the

kingdom prepared for you from the foundation of the world. (Disc) (MAGR) (INFERRED)
Contrapositive: verseF?—see 354 for similar conclusion

362. (When) . . . ye have done it unto one of the least of these my brethren, (then) ye have done it unto me (the King). (Disc) (MAMR) (INFERRED)
Contrapositive: verseF? (could have done for Jesus but not for the least)—see #372 for INVERSE

363. (When) the King shall answer and say unto them, Verily I say unto you, Inasmuch as ye have done it unto one of the least of these my brethren, (then) ye have done it unto me (the King), Then shall he (the King) say also unto them on the left hand, Depart from me, ye cursed, into everlasting fire, prepared for the devil and his angels: (Disc) (GAGR) (INFERRED)
Contrapositive: verseF?

364. (Since) I was an hungred, and ye gave me no meat . . . Then shall he (the King) say also unto them on the left hand, Depart from me, ye cursed, into everlasting fire, prepared for the devil and his angels: (Analogy-s) (MAGR) (INFERRED)
Contrapositive: verse??

365. (When) I was an hungred, (then) ye gave me no meat. (Disc) (Analogy-s) (MAMR) (INFERRED)
Contrapositive: verse? (seems to be true by definition)

366. (When) I was an thirsty, (then) ye gave me no drink. (Disc) (Analogy-s) (MAMR) (INFERRED)
Contrapositive: verse? (seems to be true by definition)

367. (When) I was a stranger, (then) ye took me not in. (Disc) (Analogy-s) (MAMR) (INFERRED)
Contrapositive: verseF?

368. (When) I was naked, (then) ye clothed me not. (Disc) (Analogy-s) (MAMR) (INFERRED)
Contrapositive: verse? (seems to be true by definition)

369. (When) I was in prison, (then) ye visited me not. (Disc) (Analogy-s) (MAMR) (INFERRED)
Contrapositive: verseF?

370. (When) the King shall say unto them on the left hand, depart from me, ye cursed . . . , Then shall they also answer him, saying, Lord, when saw we thee an hungred, or athirst, or athirst, or a stranger, or naked, or sick, or in prison, and did not minister unto thee? (Disc) GAMR) (INFERRED)
Contrapositive: verseF?

371. (When) shall they also answer him, saying, Lord, when saw we thee an hungred, or athirst, or athirst, or a stranger, or naked, or sick, or in prison, and did not minister unto thee?, Then shall he (the King) answer them saying, Verily I say unto you, Inasmuch as ye did it not to one of the least of these, ye did it not to me. (Disc) (MAGR) (INFERRED)
Contrapositive: verseF?

372. (If) Inasmuch as ye did it not to one of the least of these, (then) ye did it not to me. (Disc) (MAMR) (INFERRED)
Contrapositive: verseF? (see #362 for inverse)

373. (If one does it not unto the least of these), (then) these shall go away into everlasting punishment . . . (Disc) (MAGR) (INFERRED)
Contrapositive: verseF? (seem to imply that if go to heaven then one would also necessarily do good works)—see #374

374. After two days (have passed), (then) is the feast of the passover and the Son of man is betrayed to be crucified. (Disc) (GAGR&MR) (INFERED)
Contrapositive: verse?

375. (When) Wheresoever this gospel shall be preached in the whole world, (then) there shall also this, that this woman hath done, be told for a memorial of her. (Disc) (GAGR) (INFERRED)
Contrapositive: verse? (depends on definition of "this gospel")—see # 571 for same

376. (When one) dippeth his hand with me in the dish, (then that one) the same shall betray me. (Disc) (MAMR) (INFERRED)
Contrapositive: verseF?—see # 577 for same

377. (If) by (a man) the Son of man is betrayed, (then) woe unto that man. (Disc) (MAGR) (INFERRED)
Contrapositive: verseF?

378. If he (the man who is to betray Christ) had not been born, (then) it had been good for that man. (Disc) (GAMR) (INFERRED- THEN)
Contrapositive: verseF?

379. (Since) this is my blood of the new testament, which is shed for many for the remission of sins, (then) Drink ye all of it . . . (Disc) (GAMR) (INFERRED)
Contrapositive: verseF?

380. because of me this night (then) all ye shall be offended . . . (DISC) (GAMR) (INFERRED)
Contrapositive: verseF? (see #382) – A–>B – See # 581 in Mark for same

381. (Since) For it is written, I will smite the shepherd, and the sheep of the flock shall be scattered abroad, (then) All shall be offended because of me tonight. (Disc) (MAMR?) (INFERRED)
Contrapositive: verseF? (it is written will always be true!!)

382. . . . after I am risen again, I will go before you into Galilee. (Disc) (GAGR)
Contrapositive: verse?

383. (When in) this night, before the cock crow, (then) thou shalt deny me thrice. (Disc) (GAMR) (INFERRED)
Contrapositive: verseF? ??—see same in # 993

384. if it be possible, (then) let this cup pass from me . . . (Disc) (GAGR) (INFERRED)
Contrapositive: verse?

385. (If ye) watch and pray, (then) ye enter not into temptation. (Disc) (MAMR) (INFERRED)
Contrapositive: verse? (if "enter" means truly to act upon temptation)—see # 979 for same condition, but different conclusion

386. . . . if this cup may not pass away from me, except I drink it, (then) thy will be done. (Disc) (GAGR) (INFERRED)
Contrapositive verseF?

387. (Since) all they that take the sword shall perish with the sword, (then) put (ye) up again thy sword into his place . . . (Disc) (GAMR or MAMR) (INFERRED)
Contrapositive: verseF?

388. (if one) take the sword, (then that one) shall perish with the sword . . . (Disc) (MAGR or MAMR) (INFERRED)
Contrapositive: verseF?

389. (if) I . . . now pray to my Father, (then) . . . he shall presently give me more than twelve legions of angels. (Disc) (GAGR) (INFERRED-without if nor then in text)
Contrapositive: unknown? (A−>B)

390. (if) I . . . now pray to my Father, (and) . . . he shall presently give me more than twelve legions of angels, . . . (then) shall the scriptures be (not) fulfilled, that thus it must be. (Disc) (GAGR) (INFERRED)
Contrapositive: verse? (B−>C) See # 389

391. (When) I sat daily with you teaching in the temple, (then) ye laid no hold on me. (Disc) (GAMR) (INFERRED)
Contrapositive: verse?—see # 1000 for same

392. (When) my brethren . . . go into Galilee, . . . (then) there shall they see me. (Disc) (MAMR) (INFERRED)
Contrapositive: verse? (seems true by definition)

393. (Since) . . . All power is given unto me in heaven and earth, (then) go ye therefore, and teach all nations, baptizing them in the name of the Father, and of the Son, and of the Holy Ghost: Teaching them to observe all things whatsoever I have commanded you . . . (Disc) (GAMA) (INFERRED)
Contrapositive: verseF? (Christ's Power for Himself is not bounded by our inaction.)

CONDITIONAL/PROCEDURAL FORMS IN MARK

394. (if) you come after me, (then) I will make you to become fishers of men. (Disc) (MAGR) (INFERRED) (Inductive)
Contrapositive: verse? (see # 1 in Matthew)

395. (If we) go into the next towns, (then) I may preach there also . . . (Disc) (GAGR) (INFERRED)
Contrapositive: verseF? (without preaching can still go to town)- NOT in Matthew

396. (Since) therefore came I forth, (then) let us go into the next towns, that I may preach there also. (Disc) (GAGR) (INFERRED)
Contrapositive: verseF? (without going to next town, Christ still came forth)

397. check Mark 2: 10 (to see if parenthetical phrase is in "red")

{
398. (If one is) whole, (then that one has) no need of the physician. (Relig) (MAMR) (INFERRED)
Contrapositive: verse?—see # 75 in Matthew for same basic passage & # 598

399. (If one is) sick, (then that one has) need of the physician. (Relig) (MAMR) (INFERRED)
Contrapositive: verse? } STINV- close see # 76 in Matthew for same basic passage & # 599

{
400. . . . as long as they (the children of the bridechamber) have the bridegroom with them, (then) they cannot fast. (Relig) (Analogy- symbolic & literal) (GAMA - MAMR) (INFERRED)
Contrapositive: verseF? (could fast while bridegroom is there)—see 76a) for Inverse NOT in Matthew

401. (When) the bridegroom shall be taken away from them, and then shall they fast in those days. (Relig) (Analogy-symbolic & literal) (GAMA - MAMR) (INFERRED)
Contrapositive: verseF? } STINV-close—see # 76a for same

402. (If a) man also seweth (with a new cloth), (then) (that man) does not seweth a piece of new cloth on an old garment. (Relig) (Analogy-symbolic & literal) (GAGR – MAMR) (INFERRED)
Contrapositive: verseF? (by definition- note INFERRED though)—see Matthew

403. (Since) the new piece that filled it up taketh away from the old, and the rent is made worse, (a man does not) seweth a piece of new cloth on an old garment. (Relig) (Analogy – symbolic & literal) (GAGR – MAMR) (INFERED)
Contrapositive: verseF? (could be true, but could be false- so false?)—see Matthew

404. (If a) man putteth (wine into bottles), (then that man) does not putteth new wine into old bottles. (Relig) (Analogy-symbolic & literal) (GAGR – MAMR) (INFERRED)
Contrapositive: verseF? (by definition- note INFERRED though) – See #406 which seems to imply # 406.—see Matthew

405. (Since) the new wine doth burst the bottles, and the wine is spilled, and the bottles will be marred (then a man does not) putteth new wine into old bottles. (Relig) (Analogy – symbolic & literal) (GAGR – MAMR) (INFERED)
Contrapositive: verseF? (could be true, but could be false- so false?)—see matthew

406. (If a) man putteth (new wine into bottles), (then that man) putteth new wine into new bottles. (Relig) (Analogy-symbolic & literal) (GAGR – MAMR) (INFERRED)
Contrapositive: verseF? (by definition- note INFERRED though) See # 404 which seems to imply # 406.—see Matthew?

407. . . . when he (David) had need, and was an hungred, he, and they that were with him, (then) . . . he went into the house of God in the days of Abiathar the high priest, and did eat the shewbread, which is not lawful to eat but for the priests, and gave also to them which were with him . . . (Relig) (1 man / 1 Man - MAMR) (INFERRED) (specific to an individual)
Contrapositive: verseF? – NOT in Matthew, but is in Luke # 607

408. (since) the Sabbath was made for man, and not man for the sabbath, Therefore the Son of man is Lord also of the Sabbath. (Relig) (GAGR) (INFERRED) (Analogy)
Contrapostive: verse? (true vaccuously)- NOT in Matthew

409. . . . if a kingdom be divided against itself, (then) that kingdom cannot stand. (Relig) (MAGR) (INFERRED- though "if" was given) (Analogy)
Contrapositive: verseF?—see Matthew for similar

410. . . . if a house be divided against itself, (then) that house cannot stand. (Relig) (MAGR) (INFERRED- though "if" was given) (Analogy)
Contrapositive: verseF?—see Matthew for similar

411. . . . if Satan rise up against himself, (then) he cannot stand, but hath an end. (Relig) (MAGR) (INFERRED- though "if" was given)
Contrapositive: verseF?—see Matthew for similar

{
412. . . . (if a man) will (not) first bind the strong man, (then that man) can(not) enter into a strong man's house, and spoil his goods . . . (Relig) (MAMR) (INFERRED- interpreted "except") (Inductive)
Contrapositive: verseF? (what if strong man not there or not aware)—see Matthew for similar—see # 737 for somewhat similar, though in the negative sense

413. . . . (if a man) will first bind the strong man, . . . (then) he will spoil his house. (Relig) (MAMR) (INFERRED – interpested "except")
Contrapositive: verseF? – NOT in Matthew } STINV (close)

414. (If one commits) sins, (then the sins) shall be forgiven unto the sons of men . . . (Relig) (MAGR) (INFERRED) (Inductive)
Contrapositive: verseF? (seems to be false as "P' and "Not P" cannot both be true) – See # 119 in Matthew for somewhat similar

415. (If one commits) blasphemies wherewithsoever they shall blaspheme, (then the blasphemies shall be forgiven unto the sons of men). (Relig) (MAGR) (INFERRED) (Inductive)
Contrapositive: verseF? (seems to be false as "P' and "Not P" cannot both be true)

416. (If one) shall blaspheme against the Holy Ghost, (then that one) hath never forgiveness, but is in danger of eternal damnation . . . (Relig) (MAGR) (INFERRED)
Contrapositive: verse? , See # 46 (A—>B) for another condition in which God will not forgive (so C—> B)—see # 120 in Matthew for somewhat similar, though "speak a word" against the Holy Ghost is used in Matthew ; see # 587 for certain damnation by unbelief

417. (If one) shall do the will of God, (then) the same is my brother, and my sister, and mother. (Relig) (MAGR) (INFERRED)
Contrapositive:verse?, See # 70 in Matthew for same condition by different conclusion (A—>C) and #123 in Matthew for very similar.

418. (When) some (seed) fell by the way side, (then) . . . the fowls of the air came and devoured it up. (Mult) (Analogy – literal & and symbolic) (INFERRED) (MAGR – MAGR?)
Contrapositive: verse? – NOT in Matthew—see Luke # 669

See # 418 a) – "stony ground"—see # 670 for essentially the same

419. (When) some (seed) fell among thorns, (then) . . . the thorns grew up, and choked it, and it yielded no fruit. (Mult) (Analogy-literaliteralymbolic) (INFERRED) (MAGR – MAGR?)
Contrapositive: verse? NOT in Matthew—see # 646 for similar idea, but not same at all—see same in Luke # 671

420. (When) other (seed) fell on good ground, (then the seed) . . . did yield fruit that sprang up and increased; and brought forth, some thirty, and some sixty, and some an hundred. (Mult) (Analogy – symbolic & literal) (INFERRED) (MAGR – MAGR)
Contrapositive: verseF? (seed cannot fall to the good ground and not fall to the good ground - NOT in Matthew—see Luke # 672

421. (If one) hath ears to hear, (then) let him hear. (Mult) (GAMR) (INFERRED)
Contrapositive: verseF?—see Matthew and see Luke # 673

422. (If one is among those that were about him with the twelve), (then) unto you it is given to know the mystery of the kingdom of God . . . (Disc + Believers) (INFERRED) (GA & MA/GR)
Contrapositive: verseF? (modern day believers would seem to be able to know the mystery of God)- NOT in Matthew—see # 674 for same

423. (If one is from) them that are without, (then) all of these things are done in parables: (Disc + Believers) (INFERRED) (MAGR)
Contrapositive: verseF? (non-parables were used with those who were not without) – NOT in Matthew—see # 675

424. (Since) all these things are done in parables, (then) . . . seeing they (those that are without) may see and not perceive; and hearing they may hear, and not understand . . . (Disc + believers) INFERRED) (GAGR)
Contrapositive: verseF? (believers who perceived had parables told to them) – NOT in Matthew – See # 676

425. (If) seeing they (who are without) may see and perceive; and hearing they may hear and understand, (then) . . . at any time they should be converted, and their sins should be forgiven them. (Disc + Believers) (INFERRED) (GA&MA/GR)

Contrapositive: verseF? (they could not be converted, but they do perceive and understand- because of free will of man) – NOT in Matthew—see # 677

426. . . . when they (where the word is sown by the way side) have heard, (then) Satan cometh immediately, and taketh away the word that was sown in their hearts. (Disc + Believers) (INFERRED) (GA/God allows) – See # 127 (Mark does not have "and understandeth it not")
Contrapositive: verse? -see # 678

427. . . . when they (where the word is sown on stony ground) have heard the word, (then) immediately receive it with gladness: . . . (Disc + Believers) (GAMR) (INFERRED)
Contrapositive: verseF?—see # 679

428. (Since they, where the word is sown on stony ground) have no root in themselves, (then they) endure but for a time: . . . (Disc + Believers) (MAMR) (INFERRED)
Contrapositive: verse? – See # 128 and # 680

429. Afterward (where the word is sown on stony ground), when affliction or persecution ariseth for the word's sake, (then) immediately they are offended. (Disc + Believers) (God allows / MR) (INFERRED)
Contrapositive: verseF? – See # 128 for very similar and # 681

430. (when) they which are sown among thorns; such as hear the word. And the cares of this world, and the deceitfulness of riches, and the lusts of other things entering in, (then these things) choke the word, and it becometh unfruitful. (Disc + Believers) (God allows / MR) (INFERRED)
Contrapositive: verse? – NOT in Matthew—see # 682

431. (when) they which are sown on good ground; such as hear the word, and receive it . . . (then they) bring forth fruit, some thirtyfold, some sixty, and some an hundred. (Disc + Believers) (God allows and man receives / GR) (INFERRED)
Contrapositive: verse? – See # 130 for essentially the same, except that "understand" is used instead of "receive"—see # 683

432. (If a thing) is . . . hid, (then it) shall be manifested. (Disc + Believers) (GA or MA / GR) (INFERRED) (Inductive)
Contrapositive: verseF? – See # 87 and #349 for very similar in Matthew and in Luke # 684 for very similar

433. (If) any thing . . . was . . . kept secret, (then) it should come abroad. (Disc + Believers) (GA or MA / GR) (INFERRED)
Contrapositive: verseF? – NOT in Matthew (see # 305 and # 306 for similar condition, but different conclusion)

434. If any man have ears to hear, (then) let him hear. (Disc + Believers) (GAMR) (INFERRED)
Contrapositive: verseF? – See same in #104 and for same see # 685

435. (If) ye mete . . . with what measure . . . , (then) it shall be measured to you . . . (Disc + Believers) (MAGR) (INFERRED) (Inductive)
Contrapositive: verseF? – NOT in Matthew

436. (If one) hear(s), (then to that one) shall more be given. (Disc + Believers) (MAGR) (INFERRED) (Inductive)
Contrapositive: verseF? – Matthew has essentially similar condition, but different conclusion

{
437. (If one) hath, (then) to him shall be more given. (Disc + Believers) (MA/GA-GR and MA/GA-GR) (INFERRED)
Contrapositive: verseF? – See # 57 for different condition, but same conclusion and # 350 and # 352 for very similar

438. (If) one hath not, (then) from him shall be taken even which he hath. (Disc + Believers) (MA/GA-GR and MA-GA/GR) (INFERRED) (Inductive)
Contrapositive: verse?—see # 126 and # 351 (same) }STINV

439. . . . if a man should cast seed into the ground, and should sleep, and rise night and day, and (then) the seed should spring and grow up, (and) he knoweth not how. (Disc + Believers) (Analogy- literaliteralymbolic of kingdom of God) (MAGR) (INFERRED)
Contrapositive: verseF? – NOT in Matthew, but see #431 for similar

440. (When) the earth bringeth forth fruit of herself; (then) first the blade (is brought forth) . . . (INFERRED) (GAGR)
Contrapositive: verse? (physical phenomena)

441. (When) the earth bringeth forth fruit of herself; (and) first the blade (is brought forth), then the ear (is brought forth), after that the full corn (is) in the

ear. (Disc + Believers) (Analogy - literaliteralymbolic) (INFERRED) (GAGR)
Contrapositive: verseF? (physical phenomena) blade can come out though ear does not – NOT in Matthew

442. . . . the ear (is brought forth), after that the full corn (is) in the ear. (Disc + Believers) (Analogy- literaliteralymbolic) (INFERRED) (GAGR)
Contrapositive: verseF? (physical phenomena) – NOT in Matthew

443. . . . when the fruit is brought forth, (then) immediately he putteth in the sickle, because the harvest is come. (Disc + Believers) (Analogy-literaliteralymbolic) (GAMR-l / GAGR-s) (INFERRED) – NOT in Matthew
Contrapositive: verse? A–> B

444. . . . immediately he putteth in the sickle, because the harvest is come. (Disc + Believers) (Analogy-literaliteralymbolic) (GAGR)
Contrapositive: verse? C–>B – NOT in Matthew

445. . . . when it (a grain of mustard seed) is sown in the earth, (then it) is less than all the seeds that be in the earth . . . (Disc + Believers) (Analogy-literal) (MAGR) (INFERRED)
Contrapositive: verseF?-Not in Matthew A–>B

446. . . . when it (a grain of mustard seed) is sown, (then) it groweth up, and becometh greater than all herbs, and shooteth out great branches. (Disc + Believers) (Analogy-literal) (MAGR) (INFERED)
Contrapositive: verseF? – A–>C NOT in Matthew exactly (see # 130b) and # 447 below for essentially the same

447. when it (a grain of mustard seed) is sown, (and then) it groweth up, and becometh greater than all herbs, and shooteth out great branches, (then) the fowls of the air may lodge under the shadow of it (the tree). (Disc + Believers) (Analogy-literal) (MAGR and GAGR) (INFERED)
Contrapositive: verseF? (not necessarily the part about greater than all herbs – NOT In Matthew A–>C–>D (see # 130b) for essentially the same and # 446 above for close

(Mark 5: 34- thy 'faith' hath made thee whole (in procedural form, this does not appear to be true always for all people since one can have true 'faith' but not be made whole.) Spoken to the woman directly.—see Mark 10:51 # 514

448. (If one is) a prophet, (then that one) is not without honour, but in his own country, and among his own kin, and in his own house. (Disciples + those from Jesus' own country) (GAMR) (INFERRED)
Contrapositive: verse? – SEE #141

449. (If) . . . in what place soever ye enter into an house, (then) there abide till ye depart from that place. (Disc) (INFERRED) (MAMR)
Contrapositive: verse?) – See # 688 for essentially similar

450. (If) whosoever shall not receive you, nor hear you, (then) when ye depart thence, shake off the dust under your feet for a testimony against them. (Disc) (MAMR) (INFERRED)
Contrapositive: verseF?—see # 84 and # 689 for same and see # 96 for opposite of the condition here, but different conclusion

451. (When) . . . in the day of judgment, (then) verily it shall be more tolerable for Sodom and Gomorrha . . . than for that city (who shall not receive you, nor hear you). (Disc) (INFERRED) (GAGR)
Contrapositive: verse? – NOT in Matthew, BUT IT IS IN Luke (see # 711)

452. Be of good cheer: (since) it is I (Jesus). (Disc) (GAMR) (INFERRED)
Contrapositive: verseF? – NOT in Matthew A−>B

453. (Since) it is I (Jesus), (then) be not afraid. (Disc) (GAMR) (INFERRED)
Contrapositive: verseF? – NOT in Matthew A−>C (see # 1077 for same)

454. (Since) they (this people of Israel) teach . . . for doctrines the commandments of men, (then) in vain do they worship me . . . (Relig) (MAMR) (INFERRED)
Contrapositive: verseF? – NOT in Matthew

455. (Because) full well ye reject the commandment of God, (then) ye may keep your own tradition. (Relig) (MAMR) (INFERRED) Mark 7:9
Contrapositive: verseF? – Converse (transgress is like reject) in #142

Moses said in Mark 7:10 – "Whoso curseth father or mother, let him die the death. Religious say . . . Corban . . .

456. (When) ye suffer him (a man who says Corban to his father or mother) no more to aught for his father or mother, (then you are) making the word of

God of none effect through your tradition, which ye have delivered . . . (Relig) (MAMR) (INFERRED)
Contrapositive: verse? – See #145 for very similar
{
457. (If any) . . . thing (is) from without a man, . . . (and) enter(s) into him, (then that thing) can(not) defile him . . . (Mult) (MAGR) (INFERRED)
Contrapositive: verseF?—see #146 for same and # 155 for similar

458. (If a) . . . thing comes out of him (a man), (then that thing can) defile(s) the man . . . (Mult) (MAGR) (INFERRED)
Contrapositive: verseF? See # 147 for same and #155 for very similar}
STINV close

459. If any man have ears to hear, (then) let him hear. (Mult) (MAMR) (INFERRED)
Contrapositive: verseF?—see #104

460. Repeat of #457

461. (Any) thing from without entereth into the man, it cannot defile him, because it (anything) entereth not into his heart, but into the belly, and goeth out into the draught, purging all meats . . . (Mult) (MAMR/GR) (INFERRED)
Contrapositive: verse? – similar #150

462. Repeat of #151

463. (If) . . . evil things come from within, (then these things) . . . defile the man. (Mult) (MAGR) (INFERRED)
Contrapositive: verse? - #458 (except here the condition is more specific about things being evil)

464. (Since) it is not meet to take the children's bread, and to cast it unto the dogs, (then) let the children first be filled . . . (Greek woman) (GAGR) (INFERRED)
Contrapositive: verseF? – NOT in Matthew

465. (Since you have said) . . . this (yes, Lord: yet the dogs eat under the table eat of the children's crumbs), then go thy way; the devil is gone out of thy daughter. (Greek woman) (MAGR) (INFERRED)
Contrapositive: verseF? - NOT in Matthew

466. . . . because they have now been with me three days, and have nothing to eat, (then) I have compassion on the multitudes. (Disc) (MAGR) (INFERRED)
Contrapositive: verseF? - # 156 for same

467. . . . if I send them away fasting to their own houses, (then) they will faint by the way . . . (Disc) (GAMR) (INFERRED- though 'if' given)
Contrapositive: verseF?

468. . . . they will faint by the way: (since) divers of them came from far . . . (Disc) (GAMR) (INFERRED)
Contrapositive: verseF? – NOT in Matthew

469. . . . (since) thou savourest not the things that be of God, but the things of men, Get thee behind me, Satan . . . (Disc, incl. Peter) (MAGR) (INFERRED)
Contrapositive: verse? (vaccuously since condition of contrapositive is always met) – NOT in Matthew

470. (If one) will come after me, (then) let him deny himself, and take up his cross, and follow me. (Mult + Disc) (MAMR) (INFERRED)
Contrapositive: verse? – Same in #164 and # 690; and similar condition of this contrapositive compared to #93, but different conclusion than contrapositive here

471. (If one) will save his life, (then) he shall lose it (his life). (Mult + Disc) (MAMR) (INFERRED)
Contrapositive: verseF?—see # 94 for similar and # 165 for same

472. (If one) will lose his life for my sake and the gospel's, (then) he shall save it (his life). (Mult + Disc) (MAMR) (INFERRED)
Contrapositive: verseF?—see # 95 for similar and #166 and # 691 for same and #211 for different condition but similar condition

473. (If one) . . . therefore shall be ashamed of me and of my words in this adulterous and sinful generation, (then) of him also shall the Son of man be ashamed, when he cometh in the glory of his Father with the holy angels . . . (Disc + Mult) (MAGR) (INFERRED)
Contrapositive: verse?- NOT in Matthew – See # 692 for essentially the same

474. . . . till they (some of them that stand here) have seen the kingdom of God come with power, (then) . . . there be some of them that stand here, which shall not taste of death . . . (Mult + Disc) (GAGR) (INFERRED)
Contrapositive: verseF? – NOT in Matthew – See # 693 for very similar (note: this version has "come with power")

475. If thou canst believe, (then) all things are possible to him that believeth. (to the father of demon-possessed son) (MAGR) (INFERRED- "if" given)
Contrapositive: verse? (seems to be true by definition believe −> can believe (see # 171 for similar condition and essentially the same conclusion)—see also # 587 for similar condition, and more specific conclusion

476. (If one has the situation in which one is possessed with a demon of) "this kind", (then) this kind (of demon) can come forth (out of a person) by nothing but prayer and fasting. (Disc) (MAGR) (INFERRED)
Contrapositive: verseF? – See # 172 for essentially the same

477. . . . after that he (The Son of man) is killed, (then) he shall rise the third day . . . (Disc) (MA/GA GR) (INFERRED)
Contrapositive: verseF? - # 174 for very similar

478. If any man desire to be first, (then) the same shall be last of all, and servant of all. (Disc- The Twelve) (MAMR) (INFERRED- "if" given) (Inductive)
Contrapositive: verseF?—see # 222 for very similar

479. (If one) shall receive one of such children in my name, (then that one) receiveth me . . . (Disc- Twelve) (MAGR) (INFERRED)
Contrapositive: verseF? – See # 96 for different condition, but same conclusion and # 181 and # 695 for same as here

480. (If one) shall receive me, (then that one) receiveth not me, but him that sent me. (Disc- Twelve) (MAGR) (INFERRED)
Contrapositive: verse?—see # 479 (combined would indicate A−> B, B−>C in formal human reasoning, but not in the Gosples) # see # 695 for very similar (this version has ". . . not me . . .").—see # 1210 for essentially similar, but not the same (receive is not the same as believe on)

481. . . . (since) . . . there is no man which shall do a miracle in my name, that can lightly speak evil of me, (then) forbid him (the man which shall do a miracle in my name) not . . . (Disc) (MA-GA/MR) (INFERRED)
Contrapositive: verse? – NOT in Matthew

482. (if) one is not against us, (then that one) is on our part. (Disc) (MAMR) (INFERRED)
Contrapositive: see # 116 for very close to contrapositive, Matthew 12:30 and also see # 700 for essentially the same with slightly different phrasing but same basic meaning (can one be "neutral" about Christ?—human logic says yes, but Christ's seems to say no—not two masters)

483. . . . (since) he that is not against us is on our part, (then) there is no man which shall do a miracle in my name, that can lightly speak evil of me . . . (Disc) (MAMR) (INFERRED) (Inductive)
Contrapositive: verse? – NOT in Matthew

484. . . . (if one) shall give you a cup of water in my name, because ye belong to Christ, (then) verily I say unto you, he shall not lose his reward. (Disc) (MAGR) (INFERRED)
Contrapositive: verseF? – (see # 100 for very similar condition and same conclusion)

485. (If one) shall offend one of these little ones that believe in me, (then) it is better for him (the one that offends one of these little ones) that a millstone were hanged about his neck, and he were cast into the sea. (Disc) (MAMR) (INFERRED)
Contrapositive: verseF?—see # 182 and # 883 for same

486. If thy hand offend thee, (then) cut it off . . . (Disc) (MAMR) (Analogy) (INFERRED- "If" given)
Contrapositive: verseF?—see # 184 for same plus "foot" in addition to "hand" and "cast them from thee" in addition to "cut"; also see # 29 for same condition, very similar conclusion.

487. If thy foot offend thee, (then) cut it off . . . (Disc) (MAMR) (Analogy) (INFERRED – "IF")
Contrapositive: verseF?—see # 184 for "foot" part

488. If thine eye offend thee, (then) pluck it out . . . (Disc) (MAMR) (Analogy) (INFERRED – "IF")
Contrapositive: verseF? – See # 28 for same condition and adds "cast it off" to this verse in Mark

489. (Since) every one shall be salted with fire, and every sacrifice shall be salted with salt, (then) . . . it is better for thee to enter into the kingdom of God

with one eye, than having two eyes to be cast into hell fire, where their worm dieth not, and the fire is not quenched. (Disc) (GAGR) (Analogy- Symb) (INFERRED)
Contrapositive: verseF? – NOT in Matthew

490. (If one is a person, then that) one shall be salted with fire. (Disc) (MAGR) (Analogy) (INFERRED)
Contrapositive: verseF? – NOT in Matthew , false by definition

491. (If a thing is a sacrifice, then that) sacrifice shall be salted with salt. (Disc) (MAGR) (Analogy) (INFERRED)
Contrapositive: verseF? – NOT in Matthew, false by definition

492. (If a thing is) salt, (then that thing) is good. (Disc) (GAGR) (Analogy) (INFERRED)
Contrapositive: verse? (there are many not good things that are not salt) – See # 11 in Matthew for an inverse-like passage (though not strictly) – See #843 for same

493. . . . (because of) the hardness of your heart, (then) he (Moses) wrote you this precept (of a bill of divorcement). (Relig) (MAMR) (INFERRED)
Contrapositive: verseF? – NOT in Matthew

494. (Since) from the beginning of creation God made them male and female, (then for) this cause shall a man leave his mother and father and cleave to his wife. (Relig) (GAMR) (INFERRED)
Contrapositive: verseF? (man's inaction does not cause God's creation to be false-

495. (If) a man leave his mother and father and cleave to his wife, (then) they twain shall be one flesh . . . (Relig) (MAGR) (INFERRED)
Contrapositve: verseF? – woman could not leave or cleave—see # 199 for same

496. (If) God hath joined (a man and a woman together), (then) let not man put (the joined man and woman) asunder. (Relig) (GAMR) (INFERRED)
Contrapositve: verseF?—see # 200 for same

497. (If a man) shall put away his wife, and marry another, (then that man) committeth adultery. (Disc) (MAMR) (INFERRED)
Contrapositive: verseF?- If one does not commit adultery, he could still put his wife away for abuse or if she commits adultery.—see # 201 and #31 for

similar (though Matthew adds an additional condition as "except for fornication")

498. . . . if a woman shall put away her husband, and be married to another, (then) she committeth adultery. (Disc) (MAMR) (INFERRED)
Contrapositive: verseF?- If one does not commit adultery, she could still put her husband away for abuse or if he commits adultery. – NOT in Matthew

499. . . . (since) of such (little children) is the kingdom of God, (then) suffer the little children to come unto me, and forbid them not . . . (Disc) (GAMR) (INFERRED)
Contrapositive: verseF? – NOT in Matthew

500. (If one) shall not receive the kingdom of God as a little child, (then) he (that one which shall not receive) shall not enter therein. (Disc) (MAGR) (INFERRED)
Contrapositive: verseF? – NOT in Matthew

501. . . . (if thou) go thy way, sell whatsoever thou hast, and give to the poor, (then thou) shalt have treasure in heaven. (Disc) (MAGR) (INFERRED) (Inductive)

Contrapositive: verse?—see # 207 for same (NOTE: THIS appears to be personalized to this man, not a general statement to all)—see # 136 for "converse-like" but not totally

502. (if thou) go thy way, sell whatsoever thou hast, and give to the poor, (and thou) shalt have treasure in heaven, (then) come and follow me. (Disc) (MAGR) (INFERRED)
Contrapositive: verse?—see # 208 for similar

503. (If one) 'faith's in riches, (then) hard . . . it is for them (who 'faith' in riches) . . . to enter into the kingdom of God . . . (Disc) (MAMR?) (INFERRED)
Contrapositive: verseF?—see # 209 for similar

504. (If one only considers the possibility of a rich man being saved as) . . . men (do), (then) it (a rich man being saved) is impossible . . . (Disc) (MAMR) (INFERRED)
Contrapositive: verseF? – NOT in Matthew, but see # 171 for similar in regard to conbtrapositve, though not exactly

505. (If one only considers the possibility of a rich man being saved as) . . . men (do), (then) it (a rich man being saved) is (not) impossible with God . . . (Disc) (MAMR) (INFERRED)
Contrapositive: verseF? (true vaccuously since all things are possible with God)- NOT in Matthew

506. (If any) man . . . hath left house, or brethren, or sisters, or father, or mother, or wife, or children, or lands, for my sake, and the gospel's, (then) . . . he shall receive an hundredfold now in time, houses, and brethren, and sisters, and mothers, and children, and lands, with persecutions, and in the world to come eternal life. (Disc) (MAGR) (INFERRED)
Contrapositive: verseF? – NOT in Matthew

507. (When) . . . we (Jesus and the disciples) go up to Jerusalem, (then) the Son of man shall be delivered unto the chief priests, and unto the scribes; and they shall condemn him to death, and shall deliver him to the Gentiles: and they shall mock him, and scourge him, and shall kill him: and the third day he shall rise again. (Disc) (GA-MR&GR) (INFERRED)
Contrapositive: verseF?—see # 223 for similar

508. (If a thing) . . . is prepared (by God the Father), . . . (then) it shall be given to them . . . (Disc- John and James) (GAGR) (INFERRED)
Contrapositive: verse? – NOT in Matthew

509. (If one is) . . . accounted to rule over the Gentiles, (then that one) . . . exercise(s) lordship over them (the Gentiles). (Disc) (GAMR) (INFERRED)
Contrapositive: verseF? (depends on definition of "lordship")? – NOT in Matthew

510. (If one is of) . . . their (Gentile) great ones, (then that one) . . . exercise(s) authority upon them (the Gentiles). (Disc) (GAMR) (INFERRED)
Contrapositive: verseF? – NOT in Matthew

511. (If one) shall be your (the disciples) minister, (then that one) will be great among you . . . (Disc) (MA/GA-GR) (INFERRED- NOTE the inverted order of presentation in Mark 10:43)
Contrapositive: verse? (depends on definition of great for see #510)—see 16, 180 for different condition, but similar conclusion and # 224, # 225, # 512 for essentially the same

512. (If one) shall be servant of all, (then that one) will chiefest (among you) . . . (Disc) (MA/GA-GR) (INFERRED- NOTE the inverted order of presentation in Mark 10:43)
Contrapositive: verse?—see 16, 180 for different condition, but similar conclusion and # 224, # 225, # 511 for essentially the same (see # 990 for same)

513. (Since) even the Son of man came not to be ministered unto, but to minister, and to give his life a ransom for many, (then) whosoever of you will be the chiefest, shall be servant of all. (Disc) (GAMR) (INFERRED)
Contrapositive: verseF? – NOT in Matthew, but see # 512 just above

514. (Since you have) . . . 'faith', (then you are) . . . made . . . whole. (Disc) (MAGR) (INFERRED)
Contrapositive: verseF?—seems untrue by experience in that people with 'faith' die.- NOT in Matthew

515. . . . (When) as soon as ye be entered into it, (the village over against you), (then) ye shall find a colt tied, whereon never a man sat . . . (Disc- 2 of them) (MAGR) (INFERRED)
Contrapositive: verseF?—see # 226

516. see # 227 (only colt , not ass as well)

517. see # 228 (only colt , not ass as well)

518. see # 229 (only colt , not ass as well)

519. (If one is a human), then (that human shall not) eat fruit of thee (the fig tree with leaves but no fruit) hereafter for ever. (Disc) (MAGR) (INFERRED)
Contrapositive: verse?- would seem to be true in all cases – NOT in Matthew exactly, but a command is given in Matthew like this except without a reference to man, only to the tree instead.

520. (If one) shall say unto this mountain, Be thou removed, and be thou cast into the sea, and shall not doubt in his heart, but shall believe that those things which he saith shall come to pass, (then) he shall have whatsoever he saith. (Disc- Peter) (MAGR) (INFERRED)
Contrapositive: verse?—see # 232

521. (If) . . . ye desire (a thing), when ye pray, (and) believe that ye receive them, (then) . . . ye shall have them. (Disc-Peter) (MAGR) (INFERRED) (Inductive)

Contrapositive: verseF?- one could not receive because one prays outside of God's Will selfishly—see Luke 11:10 (# ?) for somewhat similar and #60 ; see # 233 for very similar (except no desire mentioned) – NOTE # 890 for a desire by disciples that will not be fulfilled, but note also that this condition involves praying while #890 does not directly (which is why this is not considered to be a violation).

522. . . . when ye stand praying, (then ye) forgive, if ye have aught against any. (Disc-Peter) (MAMR) (INFERRED)
Contrapositive: verseF? – not forgive but condition be met because of "free will" –see # 197 for similar but not same as # 197 does nto have condition of praying- NOT in Matthew exactly.

{
523. (If) when ye stand praying, (ye) forgive, if ye have aught against any, (then) . . . your Father also which is in heaven may forgive you your trespasses. (Disc-Peter) (MAGR) (INFERRED)- Contrapositive: verseF?- God the Father may not forgive you but you may have just not asked?—see # 45 for very similar, except no praying or "aught" against you

524. If ye do not forgive, then neither will your Father which is in heaven forgive your trespasses. (Disc) (MAGR-N and F) (INFERRED) (Inductive)- Contrapositive: verseF?} (STINV)- MAGR—see #46 for very similar; SEE # 416 for another condition that implies God will not forgive (the sin against the Holy Spirit)

525. (If) I will also ask of you (chief priests, elders, and scribes) one question, and (you) answer me, (then) I will tell you by what authority I do these things. (Relig) (GA-MA/GR) (INFERRED)
Contrapositive: verseF?—see # 234 for highly similar

526. (When) a certain man planted a vineyard, and set an hedge about it, and digged a place for the winefat, and built a tower, and let it out to husbandmen, (then the man) went into a far country. (Relig, Disc ?) (MAMR- GAGR) (Analogy- literal) (INFERRED)
Contrapositive: verseF (literal) ; ? for (symbolic)? – NOT in Matt

527. (When) . . . at the season, (then) he (the vineyard owner) sent to the husbandmen a servant, that he might receive from the husbandmen of the fruit of the vineyard. (Relig- Disc?) (GAMR- GAGR) (Analogy- literal) (INFERRED)
Contrapositive: verseF? (l); verse? (s) – See # 239 for essentially the same

528. (When) . . . they (the husbandmen) caught him (the servant), (then) . . . they beat him, and sent him away empty. (Relig-Disc?) (MAMR- MAMR) (Analogy-literal) (INFERRED)
Contrapositive: verseF? (l); verseF? (s) – See INSERT AFTER # 239 (called # 239a)

529. (When) . . . again he (the vineyard owner) sent unto them another servant, (then) . . . at him they cast stones, and wounded him in the head, and sent him away shamefully handled. (Relig-Disc?) (MAMR- GAMR) (Analogy- literal) (INFERRED)
Contrapositive: verseF? (l); verseF?—see # 240 for similar, but not exactly

530. (When) . . . again he (the vineyard owner) sent another; and him (the servant) they (the husbandmen) killed, and many others; beating some, and killing some. (Relig-Disc?) (MAMR-GAMR) (Analogy-literal) (INFERRED)
Contrapositive: verseF? (l); verseF?—see # 240 for similar, but not exactly

531. (Since) having therefore one son, his wellbeloved, (then he, the vineyard owner) sent him (the son of the vineyard owner) unto them (the husbandmen) saying, They will reverence my son. (Relig-Disc?) (MAMR – GAGR) (Analogy-literal) (INFERRED) (Inductive)
Contrapositive: verseF? (l); verse? (s) - (would seem necessarily true as conclusion is never false) – NOT in Matthew

532. (When) those husbandmen said among themselves, This is the heir; come, let us kill him, and the inheritance shall be our's, (then) they took him, and killed him, and cast him out of the vineyard. (Relig-Disc?) (MAMR-MAMR) (Analogy-literal) (INFERRED)
Contrapositive: verseF? (l); verse? (s)—see # 241 for essentially the same with slight wording difference

533. (When) he (the lord of the vineyard) will come, (then he will) destroy the husbandmen and will give the vineyard unto others. (Relig-Disc?) (MAMR-GAGR) (Analogy-literal) (INFERRED)
Contrapositive: verseF? (l); verseF? (s) – people listening said statement like this in Matthew, not Christ as in Mark

534. (If ye) bring me a penny, (then) . . . I may see it. (Relig) (MAGR) (INFERRED)
Contrapositive: verseF? (someone else could have brought it or Jesus could have seen without them bringing. – NOT in Matthew

(Note: Render to Caesar . . . and to God . . . - given as a question and a declarative, not linked by therefore as in Matthew)

535. (When) they shall rise from the dead, (then) they neither marry, nor are given in marriage; but are as the angels which are in heaven. (Relig) (GAMR) (INFERRED)
Contrapositive: verse? – NOT in Matthew, but very similar to # 944

536. (Since) He (God who spake to Moses in the bush) is not the God of the dead, but the God of the living, ye therefore do greatly err. (Relig) (God's Character- MR) (INFERRED)
Contrapositive: verseF? (God's Character is not dependent on man's actions.) – NOT in Matthew – See # 946 for very similar condition, but different conclusion

537. (Since) David himself said by the Holy Ghost, The LORD said to my Lord, Sit thou on my right hand, till I make thine enemies they footstool, David therefore himself calleth him Lord . . . (Relig-Multi-Disc?) (GA/MA-GR/MR) (INFERRED)
Contrapositive: (do not want to say either way given that this was said by the Holy Ghost unless Scripture says specifically) – NOT in Matthew

538. (If one is of) the scribes, which love to go in long clothing, and love salutations in the marketplaces, and the chief seats in the synagogues, and the uppermost rooms at feasts: which devour widows' houses, and for a pretence make long prayers, (then that one) . . . shall receive greater damnation. (Relig-Mult-Disc?) (MAGR) (INFERRED)

533. Contrapositive: verseF? (God could be merciful?)—see # 266 for essentially the same (except that this one has more conditions)—see # 757 (# 538 for essentially similar conditions plus 2 additional conditions in # 538 and somewhat similar conclusion, but not the same) – NOTE: # 539 is actually part of # 538 (which is a typographical error, but which is not planned to be changed as this would require extensive changes to references).

540. (Since) all they (the rich) did cast in of their abundance; but she (the poor widow) of her want did cast in all that she had, even all her living, (then) . . . this poor widow hath cast more in, than all they which have cast into the treasury. (Disc) (MAMR) (INFERRED)- NOTE: THIS VERSE DOES NOT MAKE SENSE LITERALLY FOR THE POOR WIDOW DID NOT GIVE LITERALLY MORE THAN THE RICH)
Contrapositive: verse? – NOT in Matthew

541. (Since) . . . many shall come in my name, saying, I am Christ; and shall deceive many, (then you) take heed lest any man deceive you. (Disc – inner circle plus Andrew) (MAMR) (INFERED)
Contrapositive: verseF? (not necessarily true it would seem)—see # 286 for essentially the same and # 304 for similar somewhat

542. See # 287 for same (rumors of wars . . .)

543. See # 288 for very similar (not come to pass, but must needs be in Mark)

544. (When) ye shall hear of wars and rumors of wars, (then) . . . the end shall not be yet. (Disc – inner circle plus Andrew) (MAGR) (INFERRED)
Contrapositive: verseF? – NOT In Matthew—see # 289 for different condition, but same conclusion

545. See # 289 for very similar ("pestilence" and "these are the beginning of sorrows" are in Mark, not in Matthew)

546. (When or since) . . . nation shall rise against nation, and kingdom against kingdom: and there shall be earthquakes in divers places, and there shall be famine and troubles: (then) these are the beginning of sorrows. (Disc-Inner circle plus Andrew) (MA-GA/GR) (INFERRED)
Contrapositive: verseF? – NOT in Matthew

547. (Since) they shall deliver you up to the councils; and in the synagogues ye shall be beaten: and ye shall be brought before rulers and kings for my sake, for a testimony against them, (then) . . . take heed to yourselves . . . (Disc – Inner circle plus Andrew) (MA-GA/MR) (INFERRED)
Contrapositive: verseF? – NOT in Matthew

548. (When) they shall deliver you up to the councils; and in the synagogues ye shall be beaten: and ye shall be brought before rulers and kings, (then this action by them is) for my sake, (and) for a testimony against them . . . (Disc – Inner circle plus Andrew) (MAGR) (INFERRED)
Contrapositive: verseF? – NOT in Matthew

549. . . . when they (rulers and kings) shall lead you, and deliver you up, (then) take no thought beforehand what ye shall speak, neither do ye premeditate . . . (Disc – inner circle plus Andrew) (MAMR) (INFERRED- when given)
Contrapositive: verseF?—see # 84a) for same

550. . . . (when) whatsoever shall be given you in that hour (when rulers and kings shall deliver you up . . .), (then) that speak ye . . . (Disc – inner circle plus Andrew) (GAMR) (INFERRED)
Contrapositive: verseF?—see 84b)

551. (since) . . . it is not ye that speak but the Holy Ghost, (then) that (whatsoever shall be given unto you in that hour) speak ye . . . (Disc – inner circle plus Andrew) (GAMR)
Contrapositive: verseF? – NOT in Matthew

552. (if) ye (are one of the disciples – at least Peter, James, John, and Andrew), (then) ye shall be hated of all men for my name's sake . . . (Disc – inner circle plus Andrew) (MAMR) (INFERRED)
Contrapositive: verse?—see # 290 for different condition, but this conclusion in part – "hate" check? (see John 14–18)

553. . . . (if one) shall endure unto the end, (then) the same shall be saved. (Disc – inner circle plus Andrew) (MAGR) (INFERRED)
Contrapositive: verse?—see # 294 for same and # 587 for same conclusion , but different condition

554. . . . when ye shall see the abomination of desolation, spoken of by Daniel the prophet, standing where it ought not, (letteth him that readeth understand), then let them that be in Judaea flee to the mountains And let him that is on the housetop not go down into the house, neither enter therein, to take any thing out of his house: And let him that is in the field not turn back again for to take up his garment. (Disc – inner circle plus Andrew) (GAMR) (INFERRED)
Contrapositive: verseF?—see # 296 through # 298

555. (if one is) . . . with child, (or one) gives suck in those days, (then) woe to them . . . (Disc – inner circle plus Andrew) (GA-MA/GR) (INFERRED)
Contrapositive: verse?—see # 300

556. (When) . . . those days (come), (then there) . . . shall be affiction, such as was not from the beginning of the creation which God created unto this time, neither shall be. (Disc – Inner circle plus Andrew) (GAGR) (INFERRED)
Contrapositive: verse?—see # 301A

557. see # 301 for same (since # 555) , then pray that flight is not in winter)

558. see # 302 for same (except the days be shortened)

559. . . . if any man shall say to you, Lo, here is Christ; or lo, he is there, (then) believe him not . . . (Disc –inner circle plus Andrew) (MAMR) (INFERRED)
Contrapositive: verseF? – essentially the same as # 303

560. essentially the same as # 304

561. same as # 309

562. same as # 311 (except that Mark adds extra condition . . .”and the powers that are in heaven shall be shaken”)

563. same as # 312

564. same as # 313 (Mark includes “uttermost part of the earth“ . . .)

565. same as # 314

566. same as # 315

567. . . . till all these things be done, (then) . . . this generation shall not pass . . . (Disc- inner circle plus Andrew) (GAGR) (INFERRED)
Contrapositive: verseF? – NOT in Matthew

568. . . . (since) ye know not when the time is, (then) take ye heed, watch and pray . . . (Disc – inner circle plus Andrew) (GAMR) (INFERRED)
Contrapositive: verseF? (would seem to be false as one could not watch and pray, but could still not know the time))—see # 319 for similar thought, though different words in condition and conclusion

569. . . . (since) ye know not when the master of the house cometh, at even, or at midnight, or at the cockcrowing, or in the morning . . . , watch ye therefore . . . (Disc – inner circle plus Andrew) (GAMR) (INFERRED - though therefore is given)
Contrapositive: verseF? – NOT in Matthew , # see 567, # 319 for similar concept, but different

570. . . . . (if the master of the house) . . . com(es) suddenly . . . (and you are not watching), (then) he (will) find you sleeping. (Disc – inner circle plus Andrew) (GA-MA/GR) (INFERRED)
Contrapositive: verse? – NOT in Matthew

571. (Since) . . . you have the poor with you always, and whensoever ye will ye may do them (the poor) good, (then) she hath wrought a good work on me. (General group) (GA-MA/MR) (INFERRED)
Contrapositive: verseF? - NOT in Matthew

572. (when) wheresoever this gospel shall be preached throughout the whole world, this also that she hath done shall be spoken of for a memorial of her. (General group) (MAMR) (INFERRED)
Contrapositive: verse?—see same in # 375

573. (when) (ye) go . . . into the city, (then) . . . there shall meet you a man bearing a pitcher of water . . . (Two disciples) (MAGR) (INFERRED)
Contrapositive: verseF? – NOT in Matthew, but see # 980 for same

574. (when) there (in the city) shall meet you a man bearing a pitcher of water, (then) follow him. (Two disciples) (MAGR) (INFERRED)
Contrapositive: verseF? – NOT in Matthew, but see # 981 for same

575. (when) wheresoever he shall go in, (then) say ye to the Goodman of the house, The Master saith, Where is the guestchamber, where I shall eat the Passover with my disciples? (Two disciples) (MAMR) (INFERRED)
Contrapositive: verseF? – NOT in Matthew, but see # 982 for same

576. . (when you say) The Master saith, Where is the guestchamber, where I shall eat the Passover with my disciples?, (then) he (the Goodman of the house) will shew you a large upper room furnished and prepared. (Two disciples) (MAMR) (INFERRED)
Contrapositive: verseF? – NOT in Matthew, but see # 983 for same

577. (when) he (the Goodman of the house) will shew you a large upper room furnished and prepared, (then) there make ready for us. (Two disciples) (MAMR) (INFERRED)
Contrapositive: verseF? – NOT in Matthew, but see # 984 for same

578. (if one) . . . is one of the twelve, that dippeth with me in the dish, (then that one shall betray me). (Disc) (MAMR) (INFERRED- Conclusion implicit)
Contrapositive: verseF?—see # 376

579. see # 377 FOR SIMILAR except that Christ added "The Son of man indeed goeth, as it is written of him" . . . and then # 377

580. . . . if he (the one who betrayed Christ) had never been born, (then) good were it for that man (who betrayed Christ) . . . (Disc) (GAMR) (INFERRED – inverted order of "if")
Contrapositive: verseF? (same in # 379)

581. . . . until that day that I drink it new in the kingdom of God, I will drink no more of the fruit of the vine . . . (Disc) (GAGR) (INFERRED- "until" order is inverted)
Contrapositive: verse? – NOT in MAtthew

582. . . . because of me . . . , All ye (the disciples) shall be offended . . . this night . . . (Disc) (GAMR) (INFERRED)
Contrapositive: verseF? – same as # 380

583. same as # 381

584. same as # 382

585. same as # 383

586. same as # 384

587. same as # 385

{
588. . . . (if one) believeth and is baptized, (then that one) shall be saved . . . (Disc – minus Judas) (MA-GA/GR) (INFERRED) (Inductive)
Contrapositive: verse?—see # 294 and # 552 for different condition, but same conclusion (see also # 475 for similar condition (except not baptism) and more general conclusion

589. . . . (if one) believeth not, (then that one) shall be damned. (Disc – minus Judas) (MAGR) (INFERRED) (Inductive)
Contrapositive: verse?—see # 416 for possible damnation from blaspheming the Holy Spirit
} STINV—see # 587 for inverse (close)

Not considering Mark 16: 17–18 as these may not be part of actual Scripture.

CONDITIONAL/PROCEDURAL FORMS FROM LUKE

590. (If one is a) man, (then that one) shall not live by bread alone. (Devil) (GAGR) (INFERRED) (Deductive)
Contrapositive: verse? – NOT in Matt/Mark

591. (If one is a) man, (then that one) shall . . . live . . . by every word of God. (Devil) (GAGR) (INFERRED) (Deductive)
Contrapositive: verseF?—see # 589 (A−> B, A −> C) NOT in Matt/Mark

592. (If one is a) . . . prophet, (then that one is) no(t) accepted in his own country. (General/Mult- Nazareth) (GAMR) (INFERRED) (Deductive)
Contrapositive: verse? – NOT in Matt or Mark

593. . . . when the heaven was shut up three years and six months, when great famine was throughout all the land, (then) many widows were in Israel in the days of Elias . . . (General/Mult-Nazareth) (GAGR) (INFERRED) (Deductive)
Contrapositive: verseF? – NOT in Matt or Mark

594. . . . when the heaven was shut up three years and six months, when great famine was throughout all the land, (then) unto none of them (the widows in Israel) was Elias sent, save Sarepta, a city of Sidon, unto a woman that was a widow. . . . (General/Mult-Nazareth) (GAGR) (INFERRED) (Deductive)
Contrapositive: verseF? – NOT in Matt or Mark

595. . . . (when) . . . in the time of Eliseus the prophet, (then) many lepers were in Israel . . . (General/Mult-Nazareth) (GAGR) (INFERRED) (Deductive)
Contrapositive: verse? – NOT in Matt or Mark

596. . . . (when) . . . in the time of Eliseus the prophet, (then) none of them (the lepers in Israel) was cleansed, saving Naaman the Syrian. (General/Mult-Nazareth) (GAGR) (INFERRED) (Deductive)
Contrapositive: verse? – NOT in Matt or Mark

597. . . . (since) therefore I am sent, (then) I must preach the kingdom of God to other cities also (in addition to Capernaum) . . . (Mult) (GAGR) (INFERRED) (Deductive)
Contrapositive: verse? – NOT in Matt or Mark

598. . . . (when) I say unto thee (the sick of the palsy), Arise, and take up thy couch, and go into thine house, (then) ye (the scribes and Pharisees) may know that the Son of man hath power upon earth to forgive sins . . . (Relig/Mult) (GAMR) (INFERRED) (Deductive)
Contrapositive: verseF? – NOT in Matt or Mark

599. See # 75 & # 398 for same

600. See # 76 & 399 for same

601. (If one is a) . . . man, (then that one) putteth a piece of a new garment upon an old . . . (Relig/Mult) (GAMR) (INFERRED) (Deductive)
Contrapositive: verseF? – NOT in matt/mark

602. see # 77 for same

603. (If one is a) . . . man, (then that one) putteth new wine into bottles . . . (Relig/Mult) (GAMR) (INFERRED) (Deductive)
Contrapositive: verseF? – NOT in Matt/Mark

604. see # 78 for same

605. see # 79 for same

606. (If one is a) . . . man (who has) drunk old wine, (then that man will not) . . . straightway desireth new (wine) . . . (Relig/Mult) (MAMR) (INFERRED) (Inductive)
Contrapositive: verseF? – NOT in Matt/Mark

607. . . . (since) he (a man) saith, The old (wine) is better, (then that man) also having drunk old wine straightway not) desireth new (wine) . . . (Relig/Mult) (MAMR) (INFERRED) (Inductive)
Contrapositive: verse? – NOT in Matt/Mark

608. . . . when himself (David) was an hungred, and they which were with him, (then) he went into the house of God, and did take and eat the shewbread, and gave also to them that were with him . . . (Relig/Mult) (1 Man A/1 Man R) (INFERRED) (NOT deductive – specific to an individual, but principle of addressing physical needs can be generalized)
Contrapositive: verseF?—see # 407 for same

609. see # 2 for same

610. see # 6 for similar (except that this passage doesn't have "thirst after righteousness")

611. (if) ye . . . weep now, (then) ye shall laugh. (Disc-Mult) (MAMR) (INFERRED) (Deductive)
Contrapositive: verseF? – Not in Matt or Mark

612. . . . when men shall hate you, and when they shall separate you from their company, and shall reproach you, and cast out your name as evil, for the Son of man's sake, (then) Rejoice ye in that day, and leap for joy . . . (Disc-Mult) (MAMR) (INFERRED) (Deductive)
Contrapositive: verseF? – Not in Matt or Mark

613. (since) . . . behold, your reward is great in heaven, (then) . . . . Rejoice ye in that day, and leap for joy . . . (Disc-Mult) (GAMR) (INFERRED) (Deductive)
Contrapositive:verseF? – NOT in Matthew or Mark – similar to # 718 (somewhat different condition and more expansive conclusion)

614. . . . (Since) in the like manner did their fathers unto the prophets, (then) behold your reward is great in heaven . . . (Disc-Mult) (MAGR) (INFERRED) (Deductive)
Contrapositive: verseF?—see 98 and 284 for "inverse like"

615. . . . (if you) . . . are rich!, (then) woe unto you! (Disc-Mult) (MAGR) (INFERRED) (Deductive) (Analogy)
Contrapositive: verseF? – NOT in Matthew or Mark

616. . . . (since) ye have received your consolation, (then) woe unto you (the rich) . . . (Disc-Mult) (MAGR) (INFERRED) (Deductive) (Analogy)
Contrapositive: verseF? – NOT in Matthew or Mark

617. . . . (if you) . . . are full!, (then) woe unto you! (Disc-Mult) (MAGR) (INFERRED) (Deductive) (Analogy)
Contrapositive: verseF? – NOT in Matthew or Mark

618. . . . (since) ye shall hunger, (then) woe unto you (the full) . . . (Disc-Mult) (MAGR) (INFERRED) (Deductive) (Analogy)
Contrapositive: verseF? – NOT in Matthew or Mark

619. . . . when all men shall speak well of you!, (then) woe unto you! (Disc-Mult) (MAGR) (INFERRED) (Deductive) (Analogy)
Contrapositive: verseF? – NOT in Matthew or Mark

620. . . . (since) . . . did their fathers (all men speaking well of you) . . . (speak well) to the false prophets, (then) woe unto you (the disciples) . . . (Disc-Mult) (MAGR) (INFERRED) (Deductive) (Analogy)
Contrapositive: verseF? – NOT in Matthew or Mark

621. . . . (if ones are) . . . your enemies, (then) love (them). (Disc-Mult) (MAMR) (INFERRED) (Deductive)
Contrapositive: verseF?—see # 39 and # 631 for same

622. . . . (if ones) . . . hate you, (then) do good to them. (Disc-Mult) (MAMR) (INFERRED) (Deductive)
Contrapositive: verseF?—see # 41 for same

623. . . . (if ones) . . . curse you, (then) bless them. (Disc-Mult) (MAMR) (INFERRED) (Deductive)
Contrapositive: verseF?—see # 40 for same

624. see # 42 for same

625. (If one) smiteth you on the one cheek, then offer him also the other (cheek). (Disc-Mult) (INFERRED) (Analogy) (Deductive) (OT-R) (MAMR)
Contrapostive: verseF? (2 diff. persons)—see # 34 for same

626. (If one) taketh away thy cloak, then forbid (him) not to take thy coat also. (Disc-Mult) (INFERRED) (Analogy) (Deductive) (OT-R) (MAMR)
Contrapositive: verseF? (2 diff. persons)—see # 35 for very similar

627. (If one) asketh of thee, then giveth to every man. (Disc-Mult) (INFERRED) (Analogy) (Deductive) (OT-R) (MAMR)
Contrapositive: verseF? (2 diff. persons)—see # 37 for same essentially

628. (If one) taketh away thy goods, then ask them not again. (Disc-Mult)) (INFERRED) (Analogy) (Deductive) (OT-R) (MAMR)
Contrapositive: verseF? (2 diff. persons) – NOT in Math or Mark

629. . . . (if) ye would that men should do to you (some thing), (then) do ye also to them (other men) likewise (the same thing). (Disc-Mult) (MAMR) (Deductive) (INFERRED)
Contrapositive: verseF? (2 diff. persons)—see # 64 for essentially the same

630. (since) if ye love them which love you, what thank have ye? . . . (and since) sinners also love those that love them, (then) as ye would that men should do to you, do ye also to them likewise. (Disc-Mult) (MAMR) (Deductive) (INFERRED)
Contrapositive: verseF? (2 different persons) – NOT in Mathew or Mark

631. (If one is a) . . . sinner, (then that one) love(s) those that love them. (Disc-Mult) (MAMR) (Deductive) (INFERRED)
Contrapositive: verseF? (Same person) – NOT in Matthew or Mark

-

632. see # 39 and # 620 for same

633. . . . (if) ye love your enemies, and do good, and lend, hoping for nothing again, (then) . . . your reward shall be great and ye shall be the children of the Highest . . . (Disc-Mult) (MAGR) (Inductive) (INFERRED)
Contrapositive: verseF? (2 different persons) – NOT in Mathew or Mark

634. . . . (since) he (the Highest) is kind unto the unthankful and evil, (then) . . . (if) ye love your enemies, and do good, and lend, hoping for nothing again, (then) . . . your reward shall be great and ye shall be the children of the Highest . . . (Disc-Mult) (GA-MA/GR) (Deductive) (INFERRED)
Contrapositive: verseF? – NOT in Matthew or Mark (see # 521 and Luke 11:10 for "inverse" like)

635. . . . (since) he (the Highest) is kind unto the unthankful and evil, (then) be ye therefore merciful, as your Father also is merciful. (Disc-Mult) (GAMR) (Deductive) (INFERRED)
Contrapositive: verseF? – Not in Matthew or Mark

636. (if ye) judge not, (then) ye shall not be judged. (Disc-Mult) (MAGR) (Inductive) (INFERRED)
Contrapositive: verseF?—see # 54 for very similar

637. (if ye) condemn not, (then) ye shall not be condemned. (Disc-Mult) (MAGR) (Inductive) (INFERRED)
Contrapositive: verseF? – NOT in Matthew or Mark

638. (if ye) forgive, (then) ye shall be forgiven. (Disc-Mult) (MAGR) (Inductive) (INFERRED)
Contrapositive: verseF?—see # 45 for very similar

639. (if ye) give, (then) it shall be given unto you. (Disc-Mult) (MAGR-MR) (Inductive) (INFERRED)
Contrapositive: verseF?—see # 57 and 437 for essentially the same

640. (if ye) give, (then) good measure, pressed down, and shaken together, and running over, shall men give into your bosom. (Disc-Mult) (MAGR) (Inductive) (INFERRED)
Contrapositive: verseF? – NOT in Matthew or Mark

641. repeat of # 435

642. (if one is a) disciple, (then that one) is not above his master . . . (Disc-Mult) (MAGR?) (INFERRED) (Deductive)
Contrapositive: verse? – NOT in Matthew or Mark

643. (If) . . . one is perfect, (then that one) shall be as his (the disciple's) master). (Disc-Mult) (MA-GA/GR) (INFERRED) (Inductive)
Contrapositive: verseF? – definition of perfect is key - NOT in Matthew or Mark

644. (if) thou hypocrite, cast out first the beam out of thine own eye, . . . (then) shalt thou see clearly to pull out the mote that is in thy brother's eye. (Disc-Mult) (MAMR) (Analogy- literal and symbolic) (Inductive) (INFERRED)
Contrapositive: verse?—see # 55 for same

{
645. If a tree is good, then (the tree) bringeth not forth corrupt fruit. (Disc-Mult) (I) (Analogy-symbolic) (None) (Deductive)
Contrapositive: verse—see # 67 for same; see # 55 for similar and see also # 69

646. If a tree is corrupt, then (the tree) bringeth not forth good fruit. (Disc-Mult) (I) (Analogy-symbolic) (None) (Deductive)
Contrapositive: verse?—see # 68 for same; see # 56 for similar, but not same
} CLOSE STINV

647. (if an object is a) tree, (then the object) is known by his own fruit. (Disc-Mult) (None)((INFERRED) (Deductive) (Analogy- literal)
Contrapositive: verse? – Not in Matthew or Mark

648. (if an object is) of thorns, (then) men do gather figs (from that object). (Disc-Mult) (None) (INFERRED) (Deductive) (Analogy- literal)
Contrapositive: verse? – NOT in Matthew or Mark

{
649. (if one is) a good man, (then that man) out of the good treasure of his heart bringeth forth that which is good. (Disc-Mult) (MAMR) (INFERRED) (Deductive)
Contrapositive: verse? (vacuously) – NOT in Matthew or Mark, but see # 909 for a statement that say "none is good, save one, that is, God"—see # 1032 for a sense of contrapositive, though not same words, but similar intent for contrapositive since "good" is not the same strictly speaking as "loved (NOT) darkness rather than light"

650. (if one is) a evil man, (then that man) out of the evil treasure of his heart bringeth forth that which is evil. (Disc-Mult) (MAMR) (INFERRED) (Deductive)
Contrapositive: verse? (vacuously) – NOT in Matthew or Mark
} CLOSE STINV—see # 1032 for a sense of converse, though not same words, but similar intent for contrapositive since "evil" is not the same strictly speaking as "loved darkness rather than light"

651. (since) of the abundance of the heart his mouth speaketh, (then either a man) out of the good treasure of his heart bringeth forth that which is good (or) out of the evil treasure of his heart bringeth forth that which is evil. (Disc-Mult) (MAMR) (INFERRED) (Deductive)
Contrapositive: verse? – NOT in Matthew or Mark

652. (if one) cometh to me, and heareth my sayings, and doeth them, (then) I will shew you to whom he is like: He is like a man which built an house, and digged deep, and laid the foundation on a rock: and when the flood arose, the stream beat vehemently upon that house and could not shake it . . . (Disc-Mult) (MAMR) (INFERRED) (Analogy) (Inductive)
Contrapositive: verse?—see # 72 for same

653. see # 73 for same basic form

654. (when) the stream did beat vehemently, and immediately it (the house built on sand) fell; and the ruin of that house was great. (Disc-Mult) (GAGR) (INFERRED) (Analogy – 1 &s)) (Deductive)
Contrapositive: verseF? – NOT in Matthew or Mark

655. (If one) shall not be offended in me, (then) blessed is he. (disciples of John the Baptist) (MAGR) (INFERRED) (Analogy-literal) (Deductive)
Contrapositive: verseF?—see # 101 for same

656. (If one) is born of women, (then that one) is not a greater prophet than John the Baptist . . . (disciples of John the Baptist) (MA/MR-GR) (INFERRED) (Deductive)
Contrapositive: verse? – NOT in Matthew or Mark

657. (if one) .. is least in the kingdom of God, (then he) is greater than he (John the Baptist). (disciples of John the Baptist) (MA-GA/GR-MR) (INFERRED) (Deductive)
Contrapositive: verseF?—see # 102 for same

658. (if one is a) .. man of this generation, (then that man is) . . . like unto children sitting in the marketplace, and calling one to another, and saying, We have piped unto you, and ye have not danced; we have mourned to you, and ye have not wept. (Relig) (GAMR) (Analogy) (INFERRED) (Deductive)
Contrapositive: verseF? – NOT in Matthew or Mark

659. (since) . . . John the Baptist came neither eating bread nor drinking wine; and ye say, He hath a devil, (then) . . . (a) man of this generation (is) . . . like unto children sitting in the marketplace, and calling one to another, and saying, We have piped unto you, and ye have not danced; we have mourned to you, and ye have not wept. (Relig) (MAMR) (Analogy) (INFERRED) (Deductive)
Contrapositive: verseF? – NOT in Matthew or Mark

660. (if a thing is) wisdom, (then it is) . . . justified of all her children. (Relig) (GAGR) (INFERRED) (Deductive)
Contrapositive: verse? – NOT in Matthew or Mark

661. . . . when they (two debtors) had nothing to pay, (then) he (the creditor) frankly forgave them both. (Simon the Pharisee- relig) (Analogy) (MAMR-literal and MAGR- symbolic) (INFERRED) (Deductive)
Contrapositive: verseF? (seems that it could definitely be false) – NOT in Matthew or Mark

662. (when) I entered into thine house, thou gavest me no water for my feet . . . (sinner woman and Simon the Pharisee) (GAMR) (INFERRED) (Specific to this instance with Simon the Pharisee and woman)
Contrapositive: verseF? – NOT in Matthew or Mark

663. (when) I entered into thine house, she hath washed my feet with tears, and wiped them with the hairs of her head . . . (sinner/woman and Simon the Pharisee) (GAMR) (INFERRED) (Specific to this instance with the woman)
Contrapositive: verseF? – NOT in Matthew or Mark

664. (when) I entered into thine house, thou gavest me no kiss . . . (sinner/woman and Simon the Pharisee) (GAMR) (INFERRED) (Specific to this instance with Simon the Pharisee and woman)
Contrapositive: verseF? – NOT in Matthew or Mark

665. (when) I entered into thine house, this woman since the time I came in hath not ceased to kiss my feet.. (sinner/woman and Simon the Pharisee) (GAMR) (INFERRED) (Specific to this instance with the woman)
Contrapositive: verseF? – NOT in Matthew or Mark

666. (when) I entered into thine house, my head with oil thou didst not anoint . . . (sinner/woman and Simon the Pharisee) (GAMR) (INFERRED) (Specific to this instance with Simon the Pharisee and woman)
Contrapositive: verseF? – NOT in Matthew or Mark

667. (when) I entered into thine house, this woman hath anointed my feet with ointment. (sinner/woman and Simon the Pharisee) (GAMR) (INFERRED) (Specific to this instance with the woman)
Contrapositive: verseF? – NOT in Matthew or Mark

668. (since) . . . she loved much, (then) Her sins, which are many, are forgiven . . . (sinner/woman and Simon the Pharisee) (MAGR) (INFERRED) (Specific to woman??)
Contrapositive: verse? – NOT in Matthew or Mark

669. (if one is) whom little is forgiven, (then) the same loveth little (sinner/woman and Simon the Pharisee) (GA-MA/MR) (INFERRED) (Deductive)
Contrapositive: verse? – NOT in Matthew or Mark

670. (when) he (the sower) sowed, (then) some (seed) fell by the way side; and it was trodden down, and the fowls of the air devoured it. (Mult) (Anal-literal) (MAGR-GAGR) (INFERRED) (Deductive)
Contrapositive: verse? (if none fell by way side, then the sower did not sow depends on soil and method of sowing which is not specified , but is assumed for listeners)

671. (when) he (the sower) sowed, (then) some (seed) fell upon a rock; and as soon as it was sprung up, it withered away, because it lacked moisture. (Mult) (Anal-literal) (MAGR-GAGR) (INFERRED) (Deductive)
Contrapositive: verseF? (if none fell upon a rock, then the sower did not sow depends on soil and method of sowing which is not specified , but is assumed for listeners)—see # 419 for same—see # 418 for same

672. (when) he (the sower) sowed, (then) some (seed) fell among thorns; and the thorns sprang up with it, and choked it. (Mult) (Anal-literal) (MAGR-GAGR) (INFERRED) (Deductive)
Contrapositive: verseF? (if none fell upon thorns, then the sower did not sow depends on soil and method of sowing which is not specified , but is assumed for listeners)—see # 419 for same

673. (when) he (the sower) sowed, (then) some (seed) fell on good ground, and bare an hundredfold. (Mult) (Anal-literal) (MAGR-GAGR) (INFERRED) (Deductive)
Contrapositive: verseF? (if none fell on good ground, then the sower did not sow depends on soil and method of sowing which is not specified , but is assumed for listeners)—see # 420 for same

674. See # 421 for same

{
675. (if one is a disciple of Christ), (then) unto you it is given to know the mysteries of the kingdom of God . . . (Disc) (MA-GA/GR) (INFERRED- conditional nature of "unto you it is given"
Contrapositive: verse?—see # 422 for same

676. (if one is not a disciple of Christ), (then) unto others (non-disciples) parables (are given) . . . (Disc) (MA-GA/GR) (INFERRED- conditional nature of "unto you it is given, . . . but to others"
Contrapositive: verse?—see # 423 for same
} like STINV

677. See # 424 for same

678. See # 425 for same

679. See # 426 for same

680. See # 427 for same

681. See # 428 for same

{
682. . . . when he (a man) hath lighted a candle, (then he) covereth it (a candle) (not) with a vessel or putteth it under a bed; . . . (MAMR) (Anal) (Disc) (Deductive)
Contrapositive: verse? - not in Matthew or Mark

681 a) . . . when he (a man) hath lighted a candle, (then he) setteth it on a candlestick, that they which enter in may see the light. (MAMR) (Anal) (Disc) (Deductive) (INFERRED)
Contrapositive: verse? - # 13 for same
} STINV

681b) (when he, a man) setteth it (a candle) on a candlestick, (then) . . . they which enter in may see the light. (Disc) (MAMR) (Anal) (Deductive) (INFERRED)
Contrapositive: verse? -

683. see # 430 for very similar (one main word different "manifest" "made known")

684. see # 429 for same

685. . . . (if) whosoever hath, (then) to him shall be given . . . (Disc) (MA-GA/GR) (Inductive) (INFERRED)
Contrapositive: verseF?—see # 436 for similar ((though 436 uses "hears" not hath, yet "hears" is in context of 684, and also # 125 and # 436 for very similar, except that ". . . . in abundance" is added there

686. . . . (if) whosoever hath not, (then) from him shall be taken even that which he seemeth to have . . . (Disc) (MA-GA/GR) (Inductive) (IN-FERRED)
Contrapositive: verseF?—see # 438 for same, except that # 685 has ("that which he seemeth to have") and also 126 and 351 for very similar

687. (if one) . . . hear(s) the word of God, and do(es) it, (then that one is) my mother and my brethren . . . (Disc) (MAGR) (Deductive) (INFERRED)

Contrapositive: verse?—see # 123 and # 417 and # 1131 for very close to same (# 686 does not have "... hear the word of God"); see also # 72 and # 651 for same condition, but different conlusion

688. ... (since) I (Jesus) perceive that virtue is gone out of me, (then) ... somebody hath touched me ... (Disc – "Peter and they that were with him") (GAMR) (Deductive) (INFERRED)
Contrapositive: verseF? – NOT in Matthew or Mark

- NOT GENERALLY APPLICABLE TO ALL: "thy 'faith' hath made thee whole" in Luke 8: 48 seems to be specific to the lady with the issue of blood, not general, though Christ did say similar things to others in other places (Neither inductive nor deductive)

- NOT GENERALLY APPLICABLE TO ALL: "fear not, believe only, and she shall be made well" to Jairus seems also specific to him. (Neither inductive nor deductive)

689. (when) ye enter into (a) house, (then) abide there, and thence depart. (Disc) (MAMR) (Deductive) (INFERRED)
Contrapositive: verse?—see # 449 for essentially similar

690. See # 84 and # 450 for same

691. See # 164 and # 470 for same

692. See # 472 and # 162 for same

693. See # 473 for very similar (Note: this version does not have "sinful and adulterous generation")

694. See # 474 for very similar (Note: this version does not have "come with power")

695. ... (since) the Son of man shall be delivered into the hands of men, (then you) let these sayings sink down into your ears ... (Disc) (GA-MA/MR) (INFERRED) (Deductive)
Contrapositive: verseF? – Seems as if the contrapositive must allow the instance where the conclusion would happen despite men's actions – NOT in Matthew or Mark

696. See # 479 and # 181 for same

697. See # 480 for very similar, but not exactly the same key words (this version does not have ". . . not me . . .")

698. . . . (since) he that is least among you all, the same shall be great, (then) whosoever shall receive me receiveth him that sent me . . . (Disc) (MAGR) (INFERRED) (Deductive)
Contrapositive: verse? – NOT in Matthew or Mark

699. (If one) . . . is least among you all, (then) the same shall be great. (Disc) (MAGR) (INFERRED) (INDUCTIVE – least among "you all" shall be great IN GENERAL)
Contrapositive: verse? – NOT in Matthew or Mark

700. . . . (since) he that is not against us is for us, (then) forbid him not (to cast out devils in Jesus' name) (Disc) (MAMR) (INFERRED) (Deductive)
Contrapositive: verse? – Not in Matthew or Mark

701. See # 482 for essentially the same and # 116 for very close contrapositive

702. . . . (since) . . . the Son of man is not come to destroy men's lives, but to save them, (then) ye know not what manner of spirit ye are of . . . (Disc) (GAMR) (INFERRED)
Contrapositive: verse? – NOT in Matthew or Mark

703. (if a) . . . man, having put his hand to the plough, look(s) back, (then that man) is (not) fit for the kingdom of God. (Disc) (MAGR) (INFERRED)
Contrapositive: verse? – NOT in Matthew or Mark

704. . . . (since) . . . the harvest truly is great, but the labourers are few, pray ye therefore the Lord of the harvest, that he would send forth labourers into his harvest. (Disc + 70 other disciples) (Anal) (GA/MA-GR) (INFERRED) (Deductive)
Contrapositive: verseF? – if one doesn't pray, then harvest can still be great – NOT in Matthew or Mark

705. . . . (when) . . . into whatsoever house ye enter, (then) first say, Peace be to this house. (Disc + 70 other disciples) (MAMR) (INFERRED) (Deductive)

Contrapositive: verseF? – if not say Peace . . . , then they could still enter – Not in Matthew or Mark

{
706. . . . if the son of peace be there, (then) your peace shall rest upon it. (Disc + 70 other disciples) (GA/MA-MR) (INFERRED) (Deductive)
Contrapositive: verseF? – NOT in Matthew or Mark

707. . . . if (the son of peace be) not (there), then it (your peace) shall turn to you again. (Disc + 70 other disciples) (GA/MA-MR) (INFERRED) (Deductive)
Contrapositive: verse?
} Close STINV – not same as # 82 and 83

708. . . . (when) . . . into whatsoever house ye enter, (then) . . . in the same house remain, eating and drinking such things as they give . . . (Disc + 70 other disciples) (MAMR) (INFERRED) (Deductive)
Contrapositive: verseF?—seems to be definitely possible for the contrapositive to be false
See # 688 and # 449 for essentially similar

709. . . . (since) . . . the labourer is worthy of his hire, (then) . . . (when) . . . into whatsoever house ye enter, . . . in the same house remain, eating and drinking such things as they give . . . (Disc + 70 other disciples) (MAMR) (INFERRED) (Deductive)
Contrapositive: verseF? – NOT in Matthew or Mark

{
710. (when) . . . whatsoever city ye enter, and they receive you, (then) eat such things as are set before you: And heal the sick that are therein, and say unto them, The kingdom of God is come nigh unto you. (Disc + 70 other disciples) (MAMR) (INFERRED) (Deductive)
Contrapositive: verseF? – more general version of # 707, but applied to “city”- NOT in Matthew or Mark—see # 114 for different condition, but part of the conclusion here (“The kingdom of God is come nigh unto you”) is in both places.

711. (when) . . . whatsoever city ye enter, and they receive you not , (then) go your ways out into the streets of the same, and say, Even the very dust of your city, which cleaveth on us, we do wipe off against you: notwithstanding

be ye sure of this, that the kingdom of God is come nigh unto you. (Disc + 70 other disciples) (MAMR) (INFERRED) (Deductive)
Contrapositive: verseF? -
}- NOT STINV, but the structure is: # 709 is A implies B and C, while # 710 is NOT A implies D and C.

712. (when) . . . that day (of judgment shall come), (then) . . . it shall be more tolerable . . . . for Sodom, than for that city (who does not receive you) . . . (Disc + 70 disciples) (GAMR) (INFERRED) (Deductive)
Contrapositive: verse? Since the day of judgment is not dependent on that city's condition.—see # 451 for basically the same

713. See # 105 for same

714. (when) . . . at the judgment, (then) . . . it shall be more tolerable for Tyre and Sidon . . . , than for you. (Disc +70 disciples) (GAMR) (INFERRED) (Deductive)
Contrapositive: verse? – NOT in Matthew or Mark

{
715. (if one) . . . heareth you (the Disciples + the 70 disciples), (then that one) heareth me; (Disc + 70 disciples) (MAMR) (INFERRED) (Deductive)
Contrapositive: verse? – NOT in Matthew or Mark

716. (if one) . . . despiseth you (the Disciples + the 70 disciples), (then that one) despiseth me; (Disc + 70 disciples) (MAMR) (INFERRED) (Deductive)
Contrapositive: verse? – NOT in Matthew or Mark
} STINV-like (not heareth is not the same as despiseth though) AND LINK to # 716 (A implies B)

717. (if one) . . . despiseth me (the Disciples + the 70 disciples), (then that one) despiseth him that sent me; (Disc + 70 disciples) (MAMR) (INFERRED) (Deductive)
Contrapositive: verseF? – Some Jewish people's beliefs seem to show the contrapositive is not true always – NOT in Matthew or Mark; NOTE: link to # 715 (B implies C)

718. (If there is a) . . . thing, (then that thing) shall (not) by any means hurt you. (Disc + 70 disciples) (GAGR) (INFERRED) (Deductive)

Contrapositive: verse?- for condition of contrapositive to occur, the conclusion of the contrapositive must be true (so it is true vaccuously) – NOT in Matthew or Mark as conditions (note that a specific form of this statement is in mark 16:18, but this has been discounted from consideration by assumption that this passage in Mark is not likely part of the original manuscripts)

719. . . . because your names are written in heaven, (then you) rejoice. (Disc + 70 disciples) (GAMR) (INFERRED) (Deductive)
Contrapositive: verseF? – having name written in heaven seems independent of one's rejoicing (though it is a commandment) – NOT in Matthew or Mark – but see # 612 for somewhat different condition, and more expansive conclusion

720. . . . (since) it seemed good in thy sight, (then) . . . I (Jesus) thank thee, O Father, Lord of heaven and earth, that thou hast hid these things from the wise and prudent, and hast revealed them unto babes . . . (Disc + 70 disciples) (GAGR) (INFERRED) (Deductive)
Contrapositive: verse? (I have to admit that the contrapositive seems as if it could be false) – NOT in Matthew or Mark

721. (If one is a) . . . thing, (then that thing is) . . . delivered to me (Jesus) of my Father: (Disc + 70 Disciples) (GAGR) (INFERRED) (Deductive-though result is specific to Jesus)
Contrapositive: verse? (would seem to be true vaccuously) – NOT in Matthew or Mark

722. (If one is a) . . . man, (then that man) knoweth (not) who the Son is, but the Father, and who the Father is, but the Son and he to whom the Son will reveal him. (Disc + 70 disciples) (GAMR) (INFERRED) (Deductive)
Contrapositive: verse? (would seem to be true vaccuously) – NOT in Matthew or Mark

723. (If one has) . . . eyes which see the things that ye see, (then those eyes) are . . . blessed (Christ's Disciples) (GAGR) (INFERRED) (Deductive)
Contrapositive: verse? – NOT in Matthew or Mark

724. (If one) . . . this do (shalt love the Lord thy God with all thy heart, and with all thy soul, and with all they strength, and with all they mind; and they neighbor as thyself), (then) . . . thou shalt live (eternally – in context of story). (Relig – one man) (MAGR) (INFERRED) (Deductive)
Contrapositive: verse? – NOT in Matthew or Mark

725. . . . when he (a certain priest) saw him (the man hurt by thieves), (then) he (the certain priest) passed by on the other side. (Disc + Relig) (MAMR) (Specific to situation – neither inductive nor deductive in literal form)
Contrapositive: verse? – NOT in Matthew or Mark

726. . . . when he (a Levite) saw him (the man hurt by thieves), (then) he (the Levite) passed by on the other side. (Disc + Relig) (MAMR) (Specific to situation – neither inductive nor deductive in literal form)
Contrapositive: verse? – NOT in Matthew or Mark

727. . . . (when) a certain Samaritan, as he journeyed, came where he (the man hurt by thieves) and when he saw him, (then) he had compassion on him, and went to him, and bound up his wounds, pouring in oil and wine, and set him on his own beast, and brought him to an inn, and took care of him. (Disc + Relig) (MAMR) (INFERRED) (Specific to situation – neither inductive nor deductive in literal form)
Contrapositive: verseF? (the Samaritan could have not done these things, but still have seen him or come to where he was) – NOT in Matthew or Mark

728. . . . on the morrow when he (the Samaritan) departed, (then) he took out two pence, and gave them to the host, and said unto him, Take care of him (the man hurt by thieves); and whatsoever thou (the host) spendest more, when I come again, I will repay thee. (Disc + Relig) (MAMR) (Specific to situation – neither inductive nor deductive in literal form)
Contrapositive: verseF? (the Samaritan could depart without giving money to host) – NOT in Matthew or Mark

729. . . . when I come again, I will repay thee. (Disc + Relig) (MAMR) (Specific to situation – neither inductive nor deductive in literal form)
Contrapositive: verseF? (the Samaritan could have come again without repaying money to host) – NOT in Matthew or Mark

730. When ye pray, (then) say, Our Father which art in heaven, Hallowed . . . . , but deliver us from evil. (Disciples) (MAMR) (Deductive)
Contrapositive: verseF? – can pray without saying these words – NOT In Matthew or Mark as procedural?

731. . . . Though he (the one awoken by visited friend) will not rise and give him (loaves), because he is his friend (the visited friend), yet because of his importunity, (then) he will rise and give him as many as he needeth. (Disciples) (MAMR) (INFERRED – then) (Deductive)
Contrapositive: verse? – NOT in Matthew or Mark

732. See # 57 for essentially same (slight wording difference)

733. See # 58 for same

734. See # 59 for same

735. (If a) . . . kingdom (is) divided against itself, (then that kingdom) is brought to desolation; (Mult) (MAGR) (INFERRED) (Deductive)
Contrapositive: verseF? – See # 110 for same

736. See # 111 for same

737. . . . if I with the finger of God cast out devils, (then) no doubt the kingdom of God is come upon you. (Mult) (GA/GR-MR) (INFERRED – then) (Deductive)
Contrapositive: verseF?—see # 114 for very similar (condition is slightly different)

738. When a man armed keepeth his palace, (then) his goods are in peace: (Mult) (MAMR) (INFERRED – then) (Deductive)
Contrapositive: verseF?—see # 115 and # 412 for somewhat similar, though not the same

739. See # 413 for similar basic meaning, except additional words.

740. See # 116 for same.

741. See # 117 for essentially the same

742. See # 122a) for essentially the same ("finding none" here, not "findeth none")

743. See # 122b) for same

744. See # 122c) for same

745. When the unclean spirit is gone out of a man, (then) he walketh through dry places, seeking rest, and finding none, he saith, I will return unto my house whence I came out. (Relig) (unclean spirit – unclean spirit) (INFERRED) (Neither inductive nor deductive)
Contrapositive: verseF? – NOT in Mathew or Mark

746. . . . (if one) hear(s) the word of God and keep(s) it, (than that one is) . . . blessed. (Mult, including specific woman who said blessed is the womb that bare Christ) (MA-GA/GR) (INFERRED) (Deductive)
Contrapositive: verse? – NOT in Matthew or Mark (though other conditions imply "blessed".

747. (since) . . . she (the queen of the south) came from the utmost parts of the earth to hear the wisdom of Solomon; and (since) behold, a greater than Solomon is here, (then) . . . the queen of the south shall rise up in the judgment with the men of this generation; and condemn them . . . (Mult) (MA-GA/MR) (INFERRED) (Specific to the queen and Christ's presence for that generation)
Contrapositive: verseF? – NOT in Matthew or Mark

748. (since) . . . they (the men of Nineve) repented at the preaching of Jonas, and, (since) behold, a greater than Solomon is here, (then) . . . the men of Nineve shall rise up in the judgment with this generation; and shall condemn it . . . (Mult) (MA-GA/MR) (INFERRED) (Specific to the men of Nineve and Christ's presence for that generation)
Contrapositive: verseF? – NOT in Matthew or Mark

749. # 681 for essentially the same (though Luke's version has "a secret place")

750. # 681a for same

751. # 681b for same

752. See # 50 for same

753. See # 51 for same

754. (since # 751 and # 752 are true) . . . therefore . . . Take heed . . . that the light which is in thee be not darkness. (Mult) (MAMR) (INFERRED) (Deductive)
Contrapositive: verseF? – whether a man takes heed, the light of the body is still eye . . .
- NOT in Mathew or Mark

755. If thy whole body therefore be full of light, having no part dark, (then) the whole shall be full of light, as when the bright shining of a candle doth give light. (Mult) (MA-GA/MR-GR) (INFERRED) (Deductive)

Contrapositive: verse? – NOT in Matthew or Mark—see # 50 for same conclusion, but different condition slightly

756. (if an object is a) . . . thing, (then that object is) . . . clean unto you. (Relig) (GAMR) (INFERRED) (Deductive)
Contrapositive: verse? – NOT in Matthew or Mark

757. (since) . . . ye tithe mint and rue and all manner of herbs, and pass over judgment and the love of God . . . , (then) . . . woe unto you, Pharisees. (Relig) (MAGR) (INFERRED) (Deductive)
Contrapositive: verse?—see # 275 for essentially the same (slightly different wording)

758. (since) . . . ye love the uppermost seats in the synagogues, and greetings in the markets, (then) . . . Woe unto you, Pharisees,,, (Relig) (MAGR) (INFERRED) (Deductive)
Contrapositive: verse? – NOT in Matthew or Mark (exactly)—see # 538 for essentially similar conditions plus 2 additional conditions in # 538 and somewhat similar conclusion, but not the same

759. (since) . . . ye are graves which appear not, and the men that walk over them are not aware of them, (then) . . . Woe unto you, Pharisees, hypocrites! (Relig) (MAGR) (INFERRED) (Deductive)
Contrapositive: verse? – NOT in Matthew or Mark

760. . . . (Since) ye lade men with burdens grievous to be borne, and ye yourselves touch not the burdens with one of your fingers, (then) . . . Woe unto you also, ye lawyers! (Relig) (MAGR) (INFERRED) (Deductive)
Contrapositive: verse? – Not in Matthew or Mark

761. (since) . . . ye build the sepulchres of the prophets, and your fathers killed them, (then) . . . Woe unto you! (lawyers) (Relig) (MAGR) (INFERRED) (Deductive)
Contrapositive: verse? – NOT in Matthew or Mark exactly, but see # 279 for the first half of the condition ("build") and the same conclusion- FORM: A and D implies C whereas # 279 is A and B implies C.—see # 762 for (A AND D implies E)

762. (since) . . . they (your fathers) indeed killed them (the prophets) , and ye build their sepulchres, (then) . . . Truly ye (lawyers!) bear witness that

ye allow the deeds of your fathers . . . (Relig) (MAGR) (INFERRED) (Deductive)
Contrapositive: verseF? – NOT in Matthew or Mark, but notice that # 761 has same conditions, but different conclusion (A AND D implies E)

763. (since) . . . they (your fathers) indeed killed them (the prophets) , and ye build their sepulchres, Therefore also said the wisdom of God, I will send them prophets and apostles, and some of them they shall slay and persecute: That the blood of all the prophets, which was shed from the foundation of the world, may be required of this generation; From the blood of Abel unto the blood of Zacharias, which perished between the altar and the temple (Relig) (MAGR) (INFERRED) (Inductive)
Contrapositive: verse?—see # 284 for very similar, but not same

764. (since) . . . ye have taken away the key of knowledge; ye entered not in yourselves, and them that were entering in ye hindered, (then) . . . Woe unto you! (lawyers) (Relig) (MAGR) (INFERRED) (Deductive)
Contrapositive: verse? – NOT in Matthew or Mark

765. see # 87 for same

766. see # 88 for same

767. . . . (since) there is nothing covered, that shall not be revealed; neither hid, that shall not be known . . . (then) Beware ye (the disciples) of the leaven of the Pharisees, which is hypocrisy . . . (Disc) (GAMR) (INFERRED) (Deductive)
Contrapositive: verseF? – NOT in Matthew or Mark

768. . . . (since) there is nothing covered, that shall not be revealed; neither hid, that shall not be known Therefore whatsoever ye have spoken in darkness shall be heard in the light; and that which ye have spoken in the ear in closets shall be proclaimed upon the housetops. (Disc) (GA/MR-GR) (INFERRED) (Deductive)
Contrapositive: verse?– NOT in Matthew or Mark

769. . . . (if one) . . . after he hath killed hath power to cast into hell; . . . (then) Fear him . . . yea, I say unto you, Fear him. (Disc) (God allows Satan/MR) (INFERRED) (Specific to Satan, but general to all men in conclusion)
Contrapositive: verseF? - # 89 for essentially the same and # 88 for close STINV

770. (since) . . . even the hairs of your head are all numbered . . . Fear not therefore, ye are of much more value than many sparrows. (Disc) (GAMR) (INFERRED) (Deductive)
Contrapositive: verseF? – NOT in Matthew or Mark

771. (if one) . . . shall confess me before men, (then) him shall the Son of man also confess before the angels of God. (Disc) (MAGR) (INFERRED) (Deductive)
Contrapositive: verseF? – See # 90 for essentially the same

772. See # 91 for essentially the same

773. See # 119 for same

774. See # 118 for same (blaspheming the Holy Ghost)

775. See # 84a) and #549 for same

776. See # 84b) and # 550 for same

777. . . . (since) a man's life consisteth not in the abundance of the things which he possesseth, (then) . . . take heed, and beware of covetousness (Disciples – "company") (MA-GA/MR) (INFERRED) (Deductive)
Contrapositive: verseF? – NOT in Matthew or Mark

778. (if one) layeth up treasure for himself, and is not rich toward God, (then) . . . God (can or shall?) say unto him, Thou fool, this night thy soul shall be required of thee . . . (Disciples – "company") (MAGR) (INFERRED) (Deductive)
Contrapositive: verse? - NOT in Matthew or Mark

779. (since) . . . (if one) layeth up treasure for himself, and is not rich toward God, (then) . . . God (can or shall?) say unto him, Thou fool, this night thy soul shall be required of thee . . . Therefore I say unto you (disciples), Take no thought for your life, what ye shall eat; neither for the body, what ye shall put on. (Disciples) (MA-GA/MR) (INFERRED) (Deductive)
Contrapositive: verseF? – NOT in Matthew or Mark

780. (since) . . . they (the ravens) neither sow nor reap; which neither have storehouse nor barn; and God feedeth them . . . (then) consider the ravens . . . (Disc) (GAMR) (INFERRED) (Deductive)
Contrapositive: verseF? – NOT in Matthew or Mark

781. See # 53 for same

782. (since) . . . all these things (eat or drink) do the nations of the world seek after: and (since) since your Father in Heaven knoweth that you have need of these things, (then) . . . seek not ye what ye shall eat, or what you shall drink, neither be ye of doubtful mind. (Disc) (MA-GA/MR) (INFERRED) (Deductive)
Contrapositive: verse? – NOT in Matthew or Mark

783. (if) . . . seek ye the kingdom of God, (then) . . . all these things shall be added unto you. (Disc) (MAGR) (INFERRED) (Inductive)
Contrapositive: verse?—see # 54a) for same

784. (since) . . . it is your Father's good pleasure to give you the kingdom, (then) . . . Fear not, little flock . . . (Disc) (GAMR) (INFERRED) (Deductive for believers)
Contrapositive: verse? – NOT in Matthew or Mark

785. (since) . . . where your treasure is, there will your heart be also, (then) . . . sell that ye have, and give alms; provide yourselves bags which wax not old, a treasure in the heavens that faileth not, where no thief approacheth, neither moth corrupteth (Disc) (GA/MR-GR) (INFERRED) (Deductive)
Contrapositive: verse? – similar to # 207 and # 501 though in reference to rich young ruler

786. Same as # 49

787. . . . when he (the lord) cometh and knocketh (when he will return from the wedding), (then) they may open unto him immediately. (Disc) (MAMR-GAMR) (Anal- literaliteralymbolic) (INFERRED) (Deductive)
Contrapositive: verse? – NOT in Matthew or Mark

788. . . . when he (the lord) cometh shall find (his servants) watching, (then) . . . blessed are those servants (and the lord) . . . shall gird himself, and make them to sit down to meat, and will come forth and serve them. (Disc) (MAMR – GA/MR-GR) (Anal- literaliteralymbolic) (INFERRED) (Deductive)
Contrapositive: verse? – NOT in Matthew or Mark—see # 789 for specific example of condition in # 788 (LOGICAL FORM; A implies C)

789. . . . if he (the lord) shall come in the second watch, or come in the third watch, and find them (the lord's servants) so (watching in those watches),

(then) blessed are those servants . . . (Disc) (MAMR – GAGR) (Anal-liter-aliteralymbolic) (INFERRED) (Deductive)
Contrapositive: verse?—see # 788 for more general - NOT in Matthew or Mark (LOGICAL FORM: B implies C, but A also implies B)

790. see # 320 for same

791. see # 321 for same

792. see # 322 for same

793. see # 324 for same

794. see # 325 for same

795. see # 326 for same

796. see # 327 for same

797. see # 328 for same

798. (When) the lord of the servant will come in a day when he looketh not for him, (then) . . . that servant, which knew his lord's will, and prepared not himself, neither did according to his will, shall be beaten with many stripes. (Disc) (MAMR – GAGR) (Anal- literaliteralymbolic) (INFERRED) (Deductive)
Contrapositive: verseF? – NOT in Matthew or Mark

799. (if a servant) . . . knew not (his lord's will), and did commit things worthy of stripes, (then that servant) . . . shall be beaten with few stripes . . . (Disc) (MAMR – MAGR) (Anal – literal) (INFERRED) (Neither inductive nor deductive)
Contrapositive: verse? – NOT in Matthew or Mark

800. (since) . . . of whomsoever much is given, of him shall much be required, (then) (When) the lord of the servant will come in a day when he looketh not for him, (then) . . . that servant, which knew his lord's will, and prepared not himself, neither did according to his will, shall be beaten with many stripes. (Disc) (MAMR – GAGR) (Anal- literaliteralymbolic) (INFERRED) (Deductive)
Contrapositive: verse? – NOT in Matthew or Mark

801. (if) . . . much is given (to one), (then) of him shall much be required . . . (Disc) (MAMR – GAGR) (Anal- literaliteralymbolic) (INFERRED) (Inductive)
Contrapositive: verseF? – NOT in Matthew or Mark

802. (since) . . . from henceforth there shall be five in one house divided, three against two, and two against three . . . The father shall be divided against the son, and the son against the father; the mother against the daughter, and the daughter against the mother; the mother in law against her daughter in law, and the daughter in law against her mother in law . . . (then Christ brings) division (on earth) . . . (Disc) (MAGR) (INFERRED) (Deductive)
Contrapositive: verseF? – because of sin there could still be division without Christ bringing it – NOT in Matthew or Mark

803. . . . When ye see a cloud rise out of the west, (then) straightway ye say, There cometh a shower . . . (Mult) (GAMR) (INFERRED) (Deductive)
Contrapositive: verseF? – NOT in Matthew or Mark the cloud can rise in the west, but this is not dependent on what a man says

804. . . . When ye see a cloud rise out of the west, (then) straightway ye say, There cometh a shower: (then) . . . so it is. (Mult) (GAMR) (INFERRED) (Deductive)
Contrapositive: verseF? – NOT in Matthew or Mark - it could not rain, but the man's view or words won't change that

805. . . . When ye see the south wind blow, (then) ye say, There will be heat . . . (Mult) (GAMR) (INFERRED) (Deductive)
Contrapositive: verseF? – NOT in Matthew or Mark

806. . . . When ye see the south wind blow, (then) ye say, There will be heat (then) it cometh to pass . . . (Mult) (GAMR) (INFERRED) (Deductive)
Contrapositive: verseF? - NOT in Matthew or Mark

807. . . . When thou goest with thine adversary to the magistrate, (then) as thou art in the way, give diligence that thou mayest be delivered from him . . . (Mult) (MA/MR-GR) (INFERRED) (Deductive)
Contrapositive: verseF?—see # 24 (more general)

808. (if one is not) . . . delivered from (the magistrate), (then) . . . he (will) . . . hale thee to the judge and the judge deliver thee to the officer, and the officer cast thee into prison. (Mult) (GAMR) (INFERRED) (Deductive)

Contrapositive: verseF?—see # 25 (not magistrate, but enemy only) - # 25 is more general case

809. see # 26 for essentially the same

810. (if) . . . thou hast not paid the very last mite, (then) . . . thou shalt not depart thence (from prison) . . . (Mult) (MA/MR-GR) (INFERRED) (Deductive)
Contrapositive: verse?—see # 26 for same

811. (if) . . . ye . . . (do not) . . . repent, (then) . . . ye shall all likewise perish (like the Galileans whose blood Pilate had mingled with their sacrifices). (Mult) (MAGR) (INFERRED) (Inductive)
Contrapositive: verseF? – NOT in Matthew or Mark

812. (since) . . . Behold, these three years I (a certain man) come seeking fruit on this fig tree, and find none: (then) cut it down. (Mult) (Anal-literal) (MAMR-GAMR?) (INFERRED) (Deductive)
Contrapositive: verseF? – NOT in Matthew or Mark

{
813. . . . if it (the fig tree) bear fruit, (then) well . . . (Mult) (Anal – literal) (GAGR & GA/MA-GR) (Deductive)
Contrapositive: verseF? – NOT in Matthew or Mark

814. . . . if (the fig tree bear) not (fruit), (then) after that thou shalt cut it down . . . (Mult) (Anal – literal) (GAGR & MAGR) (Deductive)
Contrapositive: verseF? # 69 for very similar (essentially the same, though # 69 has "and cast into the fire" after being "hewn" or "cut" down
} INVERSES

815. . . . (since) many (people), I say unto you (a man in the cities toward Jerusalem), will seek to enter in, and shall not be able, (then) Strive to enter in at the strait gate . . . (1 man directly) (MAMR) (Deductive) (INFERRED)
Contrapositive: verseF? – NOT in Matthew or Mark

816. When once the master of the house is risen up, and hath shut to the door, and ye (anyone) begin to stand without, and to knock at the door, saying, Lord, Lord, open unto us; (then) . . . and he shall answer and say unto you, I

know you not whence ye are: (Mult) (MAMR& GA-MA/GR) (Anal – literal) (Deductive) (INFERRED)
Contrapositive: verseF? – (See # 338 for similar- except that passage specifies "other virgins")

817. (When) . . . he shall answer and say unto you, I know you not whence ye are: Then shall ye begin to say, We have eaten and drunk in thy presence, and thou hast taught in our streets. (Mult) (MAMR& GAMR) (Anal – literal) (Deductive) (INFERRED)
Contrapositive: verseF? – NOT in Matthew or Mark

818. . . . when you shall see Abraham, Isaac, and Jacob, and all the prophets, in the kingdom of God, and you yourselves thrust out . . . (then) There shall be weeping and gnashing of teeth . . . (MA/GA-MR) (Mult) (Deductive) (INFERRED – order inverted in actual text)
Contrapositive: verse?—see #133 for essentially same, except that it adds "cast into . . . fire"

819. . . . (since) it cannot be that a prophet perish out of Jerusalem, (then) . . . I (Jesus) must walk today, tomorrow, and the day following . . . (Relig) (GAGR) (Condition is general, but conclusion is specific to Christ) (INFERRED – inverted order)
Contrapositive: verse? – Not In Matthew or Mark

820. . . . until the time come when ye shall say, Blessed is he that cometh in the name of the Lord, . . . Ye shall not see me . . . (Mult) (GA-MA/MR) (Specific to Jerusalem) (INFERRED – inverted order)
Contrapositive: verseF? – NOT in Matthew or Mark

821. See repeat of 285
822. When thou art bidden of any man to a wedding, (then) sit not down in the highest room (Relig) (MAMR-MAMR) (Deductive) (INFERRED)
Contrapositive: verseF? – NOT in Matthew or Mark

823. When thou art bidden of any man to a wedding, (and thou) sit . . . down in the highest room (and) a more honorable man than thou be bidden of him (the man inviting the wedding guests), (then) he that bade thee and him come and say to thee, Give this man place; and thou begin with shame to take the lowest room. (Relig) (MAMR) (Deductive) (INFERRED)
Contrapositive: verse? – NOT in Matthew or Mark

824. When thou art bidden (to a wedding), (then) go and sit down in the lowest room that when he that bade thee cometh, he may say unto thee, Friend, go up higher . . . (Relig) (MAMR) (Deductive) (INFERRED)
Contrapositive: verseF? – NOT in Matthew or Mark

825. . . . when he that bade thee cometh, he may say unto thee, Friend, go up higher . . . (Relig) (MAMR) (Deductive) (INFERRED)

826. When thou art bidden (to a wedding), go and sit down in the lowest room that when he that bade thee cometh, he may say unto thee, Friend, go up higher: then shalt thou have worship in the presence of them that sit at meat with thee. (Relig) (MAMR) (Deductive)
Contrapositive: verseF? – NOT in Matthew or Mark

827. See # 261 for same
828. See # 262 for same
829. . . When thou makest a dinner or supper, (then) call not thy friends, nor thy brethren, neither thy kinsmen, nor they rich neighbors . . . (Relig) (MAMR) (Deductive) (INFERRED)
Contrapositive: verseF? – NOT in Matthew or Mark

830. . . . When thou makest a dinner or a supper, . . . (and thou) call . . . thy friends, . . . thy brethren, . . . they kinsmen, . . . (or) they neighbours: (then) they (friends, brethren, kinsmen, or neighbours) bid thee again, and a recompence be made thee. (Relig) (MAMR) (Deductive) (INFERRED)
Contrapositive: verse? – NOT in Matthew or Mark

831. . . . when thou makest a feast, (then) call the poor, the maimed, the lame, the blind . . . (Relig) (MAMR) (Deductive) (INFERRED)
Contrapositive: verseF? – NOT in Matthew or Mark—see # 837 for specific example

832. when thou makest a feast, (and then) call the poor, the maimed, the lame, the blind . . . (then) thou shalt be blessed . . . (Relig) (MAGR) (Deductive) (INFERRED)
Contrapositive: verseF? – NOT in Matthew or Mark

833. (since) . . . they (the poor, the maimed, the lame, the blind) cannot recompense thee, (then) thou shalt be blessed . . . (Relig) (MAGR) (Deductive) (INFERRED)
Contrapositive: verseF? – NOT in Matthew or Mark

834. (since) . . . thou shalt be recompensed at the resurrection of the just, (then) thou shalt be blessed . . . (Relig) (MAGR) (Deductive) (INFERRED)
Contrapositive: verse? – NOT in Matthew or Mark

835. (if one) . . . shall eat bread in the kingdom of God, (then that one is) . . . Blessed. (Relig) (MAGR) (Deductive) (INFERRED)
Contrapositive: verseF? – NOT in Matthew or Mark

836. . . . (since) . . . all things are now ready, (then a certain man) . . . sent his servant at supper time to say to them that were bidden, Come . . . (Relig) (MAMR-GAGR) (Anal – literal) (Deductive) (INFERRED)
Contrapositive: verse? – NOT in Matthew or Mark, but a very similar parable is told of a king who destroyed those who would not come and then invited all to come in #240s

837. . . . (when) . . . that servant came, and shewed his lord these things. Then the master of the house being angry said to his servants, Go out quickly into the streets and lanes of the city, and bring in hither the poor, and the maimed, and the halt, and the blind. (Relig) (MAMR-MAGR) (Anal-literal) (Deductive) (INFERRED)
Contrapositive: verse? (Literal) - verseF? (symbolic) – NOT in Matthew or Mark (though this is a specific instance of the more general command in #831

838. (if one is of them that) . . . were bidden (to the supper and did not come), (then) none of those men . . . shall taste of my supper. (Relig) (MAMR & GA/MA-MR) (Anal – literal) (Deductive) (INFERRED)
Contrapositive: - NOT in Matthew or Mark

840. If any man come to me, and hate not his father, and mother, and wife, and children, and brethren, and sisters, yea and his own life also, (then) . . . he cannot be my disciple. (Mult) (MAMR) (Deductive) (Inferred)
Contrapositive: verse? – NOT in Matthew or Mark

840. (if one) . . . doth not bear his cross, and come after me, (then that one) cannot be my disciple. (Mult) (MA/MR-GR) (Deductive) (INFERRED)
Contrapositive: verse? – NOT in Matthew or Mark (A–>C)

841. (if one) . . . be of you that forsaketh not all that he hath, (then) he cannot be my disciple. (Mult) (MA/MR-GR) (Deductive) (INFERRED)
Contrapositive: verse? – NOT in Matthew or Mark (B–>C)

842. See # 11 for essentially similar, though this one has conclusion about "dunghill"

843. See # 492 for same

844. if the salt has lost its savour, (then) . . . men cast it out. (Mult) (GAMR - MAMR) (Anal – literal) (Deductive) (INFERRED)
Contrapositive: verseF? – NOT in Matthew or Mark

845. See # 104 for same

846. when he (a man having an hundred sheep) hath found it (the lost), (then) . . . he layeth it on his shoulders, rejoicing. (Mult) (Anal-literal) (MAMR-GAGR) (Deductive) (INFERRED)
Contrapositive: verseF?—see # 187 for very similar, except not "layeth on his shoulders"

847. . . . when he (a man having an hundred sheep) cometh home, (then) he calleth together his neighbours, saying unto them, Rejoice with me. . . . (Mult) (Anal-literal) (MAMR-GAGR) (Deductive) (INFERRED)
Contrapositive: verseF? – NOT in Matthew or Mark

848. . . . (since) I (a man having an hundred sheep) have found my sheep which was lost . . . when he (a man having an hundred sheep) cometh home, (then) he calleth together his neighbours, saying unto them, Rejoice with me. . . . (Mult) (Anal-literal) (MAMR-GAGR) (Deductive) (INFERRED)
Contrapositive: verseF? – NOT in Matthew or Mark

849. (if) . . . one sinner . . . repenteth, (then) . . . joy (like to the joy of a man who found his lost sheep) shall be in heaven . . . (Mult) (MAGR) (Inductive) (INFERRED)
Contrapositive: verseF? – NOT in Matthew or Mark

850. . . . when she hath found it (the lost piece of silver), (then) . . . she (the woman who had the ten pieces of silver) calleth her friends and her neighbours together, saying Rejoice with me . . . (Mult) (Anal-literal) (MAMR-GAGR) (Deductive) (INFERRED)
Contrapositive: verseF? – NOT in Matthew or Mark

851. (since) I (the woman who lost a silver piece) have found the piece which I had lost, (then) . . . she (the woman who had the ten pieces of silver) calleth

her friends and her neighbours together, saying Rejoice with me . . . (Mult) (Anal-literal) (MAMR-GAGR) (Deductive) (INFERRED)
Contrapositive: verseF? – NOT in Matthew or Mark

852. . . . when he (the son who left home) had spent all, (then) . . . there arose a mighty famine in that land; and he began to be in want. (Mult) (Anal-literal) (MAMR-MAMR) (Deductive) (INFERRED) (God's role in allowing the famine relative to the boy's prodigality is not certain)
Contrapositive: verseF? – NOT in Matthew or Mark

853. . . . when he came to himself, (then) . . . he said, How many hired servants of my father's have bread enough and to spare, and I perish with hunger! (Mult) (Anal-literal) (MAMR-MAMR) (Deductive) (INFERRED)
Contrapositive: verseF? – NOT in Matthew or Mark

854. . . . when he was yet a great way off, (then) . . . his father saw him, and had compassion, and ran, and fell on his neck, and kissed him. (Mult) (Anal-literal) (MAMR-MAGR) (Deductive) (INFERRED)
Contrapositive: verseF? – NOT in Matthew or Mark

855. (since) this my son was dead, and is alive again; he was lost, and is found . . . (then) the father said to his servants, Bring forth the best robe, and put it on him; and put a ring on his hand, and shoes on his feet; and bring hither the fatted calf, and kill it; and let us eat, and be merry . . . (Mult) (Anal-literal) (MAMR& MA-GA/GR) (Deductive) (INFERRED)
Contrapositive: verseF? – Not in Matthew or Mark

856. (since) . . . he (the older brother of the prodigal son) was angry, and would not go in: therefore came his father out, and entreated him. (Mult) (Anal-literal) (MAMR & MAGR) (Deductive) (INFERRED)
Contrapositive: verseF? – NOT in Matthew or Mark

857. . . . as soon as this thy son was come, which hath devoured thy living with harlots, (then) . . . thou hast killed for him the fatted calf. (Mult) (Anal-literal) (MAMR-MAGR) (Deductive) (INFERRED)
Contrapositive: verseF? – NOT in Matthew or Mark

858. . . . (since) this thy brother was dead, and is alive again; and was lost, and is found (then) . . . It was meet that we should make merry, and be glad . . . (Mult) (Anal-literal) (MAMR & MA-GA/MR-GR) (Deductive) (INFERRED)
Contrapositive: verseF? – NOT in Matthew or Mark

859. (since) . . . thou (the steward) mayest no longer be steward (then) . . . give an account of thy stewardship . . . (Mult) (Anal – literal) (MAMR & MA-GA/MR) (Deductive) (INFERRED)
Contrapositive: verseF? – NOT in Matthew or Mark

860. (if I {the steward}) . . . beg, (then) I am ashamed (Disc) (Anal-literal) (MAMR – MAMR) (Deductive) (INFERRED)
Contrapositive: verseF? – NOT in Matthew or Mark

861. . . . when I am put out of the stewardship, (then) . . . they may receive me into their houses (? the houses of those who had a debt to the king). (Disc) (Anal-literal) (MAMR – GA/MR) (Deductive) (INFERRED)
Contrapositive: verseF? – NOT in Matthew or Mark

862. . . . because he (the steward) had done wisely, (then) . . . the lord commended the unjust steward . . . (Disc) (Anal-literal) (MAMR-MAMR) (Deductive) (INFERRED)
Contrapositive: verseF? – NOT in Matthew or Mark

863. . . . (since) . . . the children of this world are in their generation wiser than the children of light . . . (then) . . . because he (the steward) had done wisely, the lord commended the unjust steward . . . (Disc) (Anal-literal) (MAMR-MAMR) (Deductive) (INFERRED)
Contrapositive: verseF? – NOT in Matthew or Mark

864. when ye fail, (then) . . . they (friends of the mammon of unrighteousness) may receive you into everlasting habitations. (Disc) (Anal-literal) (MAMR-MAMR) (Deductive) (INFERRED)
Contrapositive: verseF? – NOT in Matthew or Mark

{
865. (if one) . . . is 'faith'ful in that which is least, (then) . . . (that one) is 'faith'ful also in much . . . (Disc) (MAMR) (Deductive) (INFERRED)
Contrapositive: verseF? – NOT in Matthew or Mark – The converse also holds by use of inferred equivalency from definition used in the words attributed to Christ, ". . . he that is 'faith'ful in that which is least is 'faith'ful also in much."

866. (if one) . . . is unjust in the least, (then) . . . (that one) is unjust also in much . . . (Disc) (MAMR) (Deductive) (INFERRED)

Contrapositive: verse? – NOT in Matthew or Mark – The converse also holds by use of inferred equivalency from definition used in the words attributed to Christ, ". . . he that is unjust in that which is least is unjust also in much."
} – STINV close

867. (if one is a) .. servant, (then that servant) . . . cannot serve two masters . . . (Disc) (Anal-literal) (MAMR-MAMR) (Deductive) (INFERRED)
Contrapositive: verse? – NOT in Matthew or Mark

{
868. (since) . . . either he (a servant) will hate the one, and love the other, (then) . . . no servant can serve two masters . . . (Disc) (Anal-literal) (MAMR-MAMR) (Deductive) (INFERRED)
Contrapositive: verseF? – NOT in Matthew or Mark

869. (since) . . . either he (a servant) will hold to the one, and despise the other, (then) . . . no servant can serve two masters . . . (Disc) (Anal-literal) (MAMR-MAMR) (Deductive) (INFERRED)
Contrapositive: verseF? – NOT in Matthew or Mark
} – STINV close

870. (since) . . . that which is highly esteemed among men is abomination in the sight of God, (then) . . . ye are they which justify yourselves before men: but God knoweth your hearts . . . (Disc) (MA-GA/MR-GR) (Deductive) (INFERRED)
Contrapositive: verse? – NOT in Matthew or Mark

871. . . . until John . . . the law and the prophets were . . . (Relig) (GAGR) (Deductive) (INFERRED)
Contrapositive: verseF? – NOT in Matthew or Mark

872. . . . since that time (the time of John), (then) . . . the kingdom of God is preached, and every man presseth into it. (Relig) (MA-GA/MR) (Deductive) (INFERRED)
Contrapositive: verse? – NOT in Matthew or Mark

873. (if one is a) . . . man, (than that one) . . . presseth into it (the kingdom of God). (Relig) (GAMR) (Deductive) (INFERRED)
Contrapositive: verse? – NOT in Matthew or Mark

874. (if one) . . . putteth away his wife, and marrieth another, (then that one) committeth adultery . . . (Relig) (MAMR) (Deductive) (INFERRED)

Contrapositive: verseF? – people can be not in adultery and never married—see # 497 and # 498 for same and # 201 for same plus adding condition of fornication

875. (if one) . . . marrieth her that is put away from her husband, (than that one) committeth adultery. (Relig) (MAMR) (Deductive) (INFERRED)
Contrapositive: verseF? – same as # 874; see # 32 and # 202 for same

876. (since) . . . I (the rich man) am tormented in this flame, (then) Father Abraham, have mercy on me, and send Lazarus, that he may dip the tip of his finger in water, and cool my tongue . . . (Relig) (GA-MA/MR) (Deductive) (INFERRED)
Contrapositive: verseF? – NOT in Matthew or Mark

877. (since) . . . between us (Abraham and Lazarus) and you (the rich man) there is a great gulf fixed, (then) . . . they which would pass from hence to you cannot; neither can they pass to us, that would come from thence. (Relig) (GA/MR or anyone) (Deductive) (INFERRED)
Contrapositive: verseF? – NOT in Matthew or Mark

878. (since 877 is true), . . . I (the rich man) pray thee, therefore, father (Abraham) that thou wouldest send him (Lazarus?) to my father's house. (Relig) (GA-MA/MR) (Deductive) (INFERRED)
Contrapositive: verseF? – NOT in Matthew or Mark (not sure if #877 is conditional for #878, but the use of "therefore" seems to suggest there is a connection)

879. (since) . . . I (the rich man) have five brethren, (then) . . . I pray thee (Abraham) therefore, father, that thou wouldest send him to my father's house . . . that he may testify unto them (the rich man's family); lest they also come into this place of torment. (Relig) (MAMR) (Deductive) (INFERRED)
Contrapositive: verseF? – NOT in Matthew or Mark

880. . . . if one went unto them from the dead, (then) . . . they will repent. (Relig) (GAMR) (Deductive) (INFERRED- BUT "IF" INCLUDED)
Contrapositive: verseF? – NOT in Matthew or Mark

881. . . . If they hear not Moses and the prophets, (then) . . . neither will they be persuaded, though one rose from the dead. (Relig) (MA/MR-GR) (Inductive) (INFERRED- though IF given)
Contrapositive: verse? – NOT in Matthew or Mark (see # 1074 for close STINV)

882. (If one is a person) . . . through whom . . . (offences) . . . come, (then) . . . woe unto him . . . (Disc) (MAGR) (Deductive) (INFERRED)
Contrapositive: verseF?—see # 183 for same

883. see # 182 and # 485 for same

884. . . . If thy brother trespass against thee, (then you) rebuke him . . . (Disc) (MAMR) (Deductive) (INFERRED-though IF given)
Contrapositive: verseF? – NOT in Matthew or Mark

885. . . . if thy brother repents, (then you) forgive him . . . (Disc) (MAMR) (Deductive) (INFDERRED-though IF given)
Contrapositive: verseF? – out forgiveness does not prevent anyone from repenting. NOT in Matthew or Mark

886. . . . if he (thy brother) trespass against thee seven times in a day, and seven times in a day turn again to thee, saying, I repent, (then) . . . thou shalt forgive him. (Disc) (MAMR) (Inductive) (INFERRED- though IF given)
Contrapositive: verseF? – NOT in Matthew or Mark; though similar to #197

887. If ye had 'faith' as a grain of mustard seed, (then) . . . ye might say unto this sycamine tree, Be thou plucked up by the root, and be thou planted in the sea; and it should obey you. (Disc) (MA-GA/MR-GR) (INFERRED- though IF given) (Deductive)
Contrapositive: verseF? – NOT in Matthew or Mark

888. . . . when ye shall have done all those things which are commanded, (then) . . . say, We are unprofitable servants; we have done that which was our duty to do. (Disc) (MAMR) (INFERRED) (Deductive)
Contrapositive: verseF? – NOT in Matthew or Mark

889. (since) . . . behold the kingdom of God is within you, (then) . . . Neither shall they say, Lo here! or lo there! . . . (Relig) (GA-MA/MR) (INFERRED) (Deductive)
Contrapositive: verseF? - NOT in Matthew or Mark

890. . . . when ye shall desire to see one of the days of the Son of man, (then) . . . ye shall not see him (Disc) (MA/GR-MR) (INFERRED) (Deductive)
Contrapositive: verseF? – NOT in Matthew or Mark – See though # 521 for an instance in which desire plus belief by disciples is to be "received by them, but note also that this condition involves praying while #890 does not directly (which is why this is not considered to be a violation).

891. (since) . . . as the lightning, that lighteneth out of the one part under heaven, shineth unto the other part under heaven; (then) . . . go not after them (who shall say) . . . See here; or see there . . . (Disc) (GAMR) (INFERRED) (Deductive)
Contrapositive: verseF? - – See # 306 for the same for essentially same, except that the conclusion in # 891 does not have the "secret chamber" portion.

892. (After) . . . he (the Son of man) . . . must . . . suffer many things, and be rejected of this generation, (then) . . . the Son of man . . . shall . . . be (as the lightning . . . under heaven) . . . (Disc) (GA-MA/GR) (INFERRED) (Neither deductive nor inductive as this is specific to Christ)
Contrapositive: verse? – NOT in Matthew or Mark

893. . . . until the day that Noe entered into the Ark and the flood came, and destroyed them all, (then) . . . they did eat, they drank, they married wives, they were given in marriage . . . (Disc) (MA-GA/MR) (INFERRED) (Deductive)
Contrapositive: verseF? – NOT in Matthew or Mark

894. . . . when the Son of man is revealed (then) . . . shall it be . . . (as with those in the days of Noe and Lot). (Disc) (GAMR) (INFERRED) (Deductive)
Contrapositive: verse? – NOT in Matthew or Mark

895. (when) . . . in that day (that the Son of man is revealed), (then) . . . he which shall be upon the housetop, and his stuff in the house, let him not come down to take it away (Disc) (GAMR) (INFERRED) (Deductive)
Contrapositive: verseF? – similar item in # 298 – condition is slightly different in that it talks about the "abomination of desolation"

896. (when) . . . in that day (that the Son of man is revealed), (then) . . . he that is in the field, let him likewise (to man on housetop) not return back (Disc) (GAMR) (INFERRED) (Deductive)
Contrapositive: verseF? - – similar item in # 298 – condition is slightly different in that it talks about the "abomination of desolation"

897. see # 165 and # 471 for same

898. see # 166 and # 472 for same

899. . . . in that night (when the Son of man is revealed and when) . . . there shall be two men in one bed; (then) . . . one shall be taken, and the other shall be left. (Disc) (GA/GR-MR) (INFERRED) (Deductive)

Contrapositive: verseF? – NOT in Matthew or Mark

900. See # 317 for same

901. See # 316 for same

902. See # 307 for same

903. . . . afterward he (the judge that feared not God neither regarded man) said within himself, Though I fear not God nor regard man, yet because of this widow troubleth me, (then) . . . I will avenge her (the widow), lest by her continual coming she weary me. (Disc) (MAMR) (Anal-literal, but the "judge" is not God as God does not think this way) (INFERRED) (Neither deductive nor inductive)
Contrapositive: verseF? – NOT in Matthew or Mark

904. (if I, the judge who fears not God nor regards man, do not) avenge her (the widow), (then).. by her continual coming she (will) weary) me. (Disc) (MAMR) (Anal- literal, but "judge" is not God as God does not think this way) (INFERRED) (Inductive)
Contrapositive: verseF? – NOT in Matthew or Mark

905. See # 261 and # 827 for same

906. See # 262 and # 828 for same

907. See # 499 for same

908. See # 500 for same

909. (if one is a person), (then that one) .. is not good, save one, that is, God. (Rich young ruler) (GA/MR-GR) (INFERRED) (Deductive)
Contrapositive: verse? – See # 649 for a condition which uses the same Greek word for "good" to describe a man

910. Same as # 207 and # 501

911. Same as # 209

912. Same as # 504

913. Essentially similar to # 506 ("manifold" here vs. "hundredfold")

914. Essentially similar to # 507, but this statement has "written by the prophets concerning the Son of man shall be accomplished", but does not have "chief priests" as #507 does

915. (if you have) . . . 'faith', (then you will be) . . . saved . . . (blind man) (MAGR) (INFERRED) (Inductive)
Contrapositive: verse?

916. (since) . . . today I must abide at thy (Zacchaeus') house (then) . . . Zacchaeus, make haste, and come down (from the sycomore tree) . . . (Zacchaeus) (GAMR) (INFERRED) (Neither inductive nor deductive)
Contrapositive: verseF? – NOT in Matthew or Mark

917. (since) . . . he (Zacchaeus) is also is a son of Abraham, (then) . . . This day is salvation come to this house . . . (Zacchaeus) (GAGR) (INFERRED) (Neither inductive nor deductive)
Contrapositive: verseF? – NOT in Matthew or Mark

918. (since) the Son of man is come to seek and to save that which was lost, (then) . . . This day is salvation come to this house . . . (Zacchaeus) (GAGR) (INFERRED) (Neither inductive nor deductive)
Contrapositive: verseF? – NOT in Matthew or Mark

919. . . . when he (the nobleman) was returned, having received the kingdom, then he commanded his servants to be called unto him, to whom he had given the money, that he might know how much every man had gained by trading. (Mult) (Anal-literal) (MAMR-GAGR) (Deductive)
Contrapositive: verseF? – similar to # 344

920. Very Similar to # 345

921. Very Similar to # 346

922. Very Similar to # 347

923. Very Similar to # 348

924. Very Similar to # 349

925. Very Similar to # 350

926. Same as # 351

927. (if one is of) . . . those mine enemies, which would not that I should reign over them, (then) . . . bring them hither, and slay them before me. (Mult) (Anal- literal) (MAMR-MAGR) (INFERRED) (Deductive)
Contrapositive: verse? - NOT in Matthew or Mark

928. Same in # 515

929. Same in # 516

930. Same in # 517

931. . . . if these (the multitude) should hold their peace, (then) . . . the stones should immediately cry out. (Relig) (MAGR) (INFERRED- though IF given) (Neither inductive nor deductive)
Contrapositive: verse? – NOT in Matthew or Mark

932. . . . (since) . . . thou (the city of Jerusalem) hadst (NOT) known, even thou, at least in this thy day, the things which belong unto they peace!, (then these things that belong unto they peace) . . . are hid from thine eyes. (City of Jerusalem) (MAGR) (INFERRED) (Neither inductive nor deductive)
Contrapositive: verseF? – NOT in Matthew or Mark

933. (since these things that belong unto thy peace are hid from thine eyes, then) . . . the days shall come upon thee (Jerusalem), that thine enemies shall cast a trench about thee, and compass thee round, and keep thee in on every side, And shall lay thee even with the ground, and they children within thee; and they shall not leave in thee one stone upon another . . . (City of Jerusalem) (GA/GR-MR) (INFERRED) (Neither inductive nor deductive)
Contrapositive: verse? – NOT in Matthew or Mark – though the section on “not one stone upon another” is in both Matthew and Mark

934. . . . because thou (Jerusalem) knewest not the time of thy visitation (then) . . . the days shall come upon thee (Jerusalem), that thine enemies shall cast a trench about thee, and compass thee round, and keep thee in on every side, And shall lay thee even with the ground, and they children within thee; and they shall not leave in thee one stone upon another . . . (City of Jerusalem) (MA/GR-MR) (INFERRED) (Neither inductive nor deductive)

Contrapositive: verse? - NOT in Matthew or Mark – though the section on "not one stone upon another" is in both Matthew and Mark

935. (when) . . . at the season (of the harvest), (then) . . . he (a certain man) sent a servant to the husbandmen, that they should give him of the fruit of the vineyard . . . (Mult) (Anal-literal) (GAMR-GAGR) (INFERRED) (Deductive)
Contrapositive: verseF? – same in # 527

936. Very similar in # 529

937. Very similar in # 530

938. Very similar in # 531

939. Same in # 532

940. Very similar in # 533

941. (if one) . . . shall fall upon that stone (the one which the builders rejected), (then that one) . . . shall (be) broken; (Mult) (Anal-s) (MAMR-MAGR) (INFERRED) (Deductive)
Contrapositive: verse? (symbolic)—see # 242 for same

942. See # 243 for same

943. See # 253 for same

944. (if one is of the) . . . children of this world, (that that one may) marry, and are given in marriage (Mult) (MAMR) (INFERRED) (Deductive)
Contrapositive: verseF? – depends on the meaning of "children of this world", but I take this to mean those who are not believers in which case this would be verseF? , but otherwise verse?

945. Similar to # 535, but # 535 is a subset of this one with one more condition and one more conclusion

946. (since) . . . he (God) is not a God of the dead, but of the living . . . (then) . . . Now that the dead are raised, even Moses shewed at the bush, when he calleth the Lord the God of Abraham, and the God of Isaac, and the God of Jacob. (Relig) (GA/GR-MR) (INFERRED) (Deductive)

Contrapositive: verseF?—see # 536 for very similar condition, but different conclusion.

947. (since) . . . all live unto him (God), (then) .. he (God) is not a God of the dead, but of the living . . . (Relig) (MAGR) (INFERRED) (deductive)
Contrapositive: verse? – Note linkage of this as a form of A−>C and B−>A

948. See # 537 for same

949. See # 538–539 for same

950. See # 540 for essentially the same (though Luke adds "out of her penury")

951. See # 541 for same

952. (since many shall come in my name, but are deceivers), . . . go ye not therefore not after them. (Mult) (MAMR) (NOT INFERRED) (Deductive)
Contrapositive: verseF? – NOT in Matthew or Mark

953. See # 542 for same

954. See # 543 for same

955. See # 544 for same

956. See # 548 for same

957. See # 549 for same

958. . . . before all these things (nation shall rise against nation, . . .) they shall lay their hands on you, persecute you, delivering you up to the synagogues, and into prisons, being brought before kings and rulers for my name's sake . . . (Mult) (GA-MA/MR) (NOT INFERRED) (Specific to disciples and/or possibly general to all believers)
Contrapositive: verse? – NOT in Matthew or Mark

959. (since) . . . I (Christ) will give you a mouth and wisdom, which all your adversaries shall not be able to gainsay nor resist, (then) . . . Settle it therefore in your hearts, not to meditate before what ye shall answer. (Mult) (GAMR) (INFERRED) (Deductive)

Contrapositive: verse? – NOT in Matthew or Mark

960. See # 552 for same

961. . . . when ye shall see Jerusalem compassed with armies, then know that the desolation thereof is nigh (Mult) (MA/GA-MR/GR) (NOT INFERRED) (Deductive)
Contrapositive: verseF? – NOT in Matthew or Mark

962. see # 297 for same

963. . . . (since) these be the days of vengeance, that all things might be fulfilled, Then let them which are in Judaea flee to the mountains, and let them which are in the midst of it depart out; and let not them that are in the countries enter thereinto. (Mult) (GAMR) (INFERRED) (Deductive)
Contrapositive: verseF? - NOT in Matthew or Mark

964. see # 299 for same

965. see # 300 for same

966. . . . (since) there shall be great distress in the land, and wrath upon this people . . . (then) . . . woe unto them that are with child, and to them that give suck, in those days! (Mult) (GAMR) (INFERRED) (Deductive)
Contrapositive: verseF? – NOT in Matthew or Mark

967. . . . until the times of the Gentiles be fulfilled . . . they (Jerusalem) shall fall by the sword, and shall be led away captive into all nations and Jerusalem shall be trodden down of the Gentiles (Mult) (GAMR) (NOT INFERRED) (Deductive)
Contrapositive: verseF? – NOT in Matthew or Mark

968. (since men will have) . . . fear, (then) men's hearts (will be) . . . failing them . . . (Mult) (MAMR) (INFERRED) (Deductive)
Contrapositive: verse? – NOT in Matthew or Mark

969. (since) . . . the powers of heaven shall be shaken, (then) . . . Men's hearts (will be) failing them, and looking after those things which are coming on the earth . . . (Mult) (GAMR) (INFERRED) (Deductive)
Contrapositive: verseF? – NOT in Matthew or Mark

970. (when) . . . the powers of heaven shall be shaken, (then) . . . shall they see the Son of man coming in a cloud with power and great glory. (Mult) (GAMR) (INFERRED) (Deductive)
Contrapositive: verseF? – NOT in Matthew or Mark

971. . . . when these things begin to come to pass, then look up, and lift up your heads . . . (Mult) (GAMR) (NOT INFERRED) (Deductive)
Contrapositive: verseF? – NOT in Matthew or Mark

972. (since) . . . your redemption draweth nigh . . . , then look up, and lift up your heads . . . (Mult) (GA-MA/MR) (INFERRED) (Deductive)
Contrapositive: verseF? – NOT in Matthew or Mark

973. When they (the trees) now shoot forth, (then) . . . ye see and know of your own selves that summer is now nigh at hand. (Mult) (MAMR) (INFERRED) (Deductive)
Contrapositive: verseF?—see # 314 for essentially the same though condition is stated with different words that have same effect

974. . . . when ye see these things come to pass, (then) . . . know ye that the kingdom of God is nigh at hand. (Mult) (MA/MR-GR) (INFERRED) (Deductive)
Contrapositive: verseF? – These things can (and I think will) happen whether we know the kingdom of God is near or not. – NOT in Matthew or Mark

975. . . . (un)til all be fulfilled, . . . This generation shall not pass away . . . (Mult) (GA/GR-MR) (INFERRED) (Deductive)
Contrapositive: verseF? – NOT in Matthew or Mark

976. (if you don't) . . . take heed to yourselves . . . (and if) . . . at any time your hearts be overcharged with surfeiting, and drunkenness, and cares of this life, (then) . . . that day (when all is fulfilled) come upon you unawares . . . (Mult) (MA/MR-GR) (INFERRED) (Deductive)
Contrapositive: verse? – NOT in Matthew or Mark

977. (since) . . . as a snare shall it (the day when all if fulfilled) come on all them that dwell on the face of the whole earth . . . (then) take heed to yourselves . . . (Mult) (GAMR) (INFERRED) (Deductive)
Contrapositive: verseF? – If we don't take care to ourselves, then these will still happen independent of man's actions – NOT in Matthew or Mark

978. (since) . . . as a snare shall it (the day when all is fulfilled) come on all them that dwell on the face of the whole earth. Watch ye therefore, and pray always . . . (Mult) (GAMR) (INFERRED) (Deductive)
Contrapositive: verseF? – If we don't watch or don't pray, then condition will still occur – NOT in Matthew or Mark

979. (if ye) . . . watch . . . and pray always, (then) . . . ye may be accounted worthy to escape all these things that shall come to pass, and to stand before the Son of man. (Mult) (MAGR) (INFERRED) (Deductive)
Contrapositive: verseF? – See # 385 for same condition, but different conclusion – NOT in Matthew or Mark

980. see # 573 for same

981. see # 574 for same

982. see # 575 for same

983. see # 576 for same

984. see # 577 for same

985. (since) . . . I say unto you, I will not any more eat thereof, until it (the passover) be fulfilled in the kingdom of God, (then) . . . With desire I have desired to eat this passover with you before I suffer. (Disc) (GAGR) (INFERRED) (Neither inductive nor deductive)
Contrapositive: verse? – NOT in Matthew or Mark

986. . . . until it (the passover) be fulfilled in the kingdom of God, . . . I say unto you, I will not any more eat thereof . . . (Disc) (GAGR) (INFERRED) (Neither inductive nor deductive)
Contrapositive: verse? – NOT in Matthew or Mark

987. (since) . . . I say unto you, I will not drink of the vine, until the kingdom of God shall come . . . (then) . . . Take this, and divide it (drink of the vine) among yourselves . . . (Disc) (GAMR) (INFERRED) (Neither inductive nor deductive)
Contrapositive: verseF? – NOT in Matthew or Mark

988. . . . until the kingdom of God shall come . . . I say unto you, I will not drink of the vine . . . (Disc) (GAGR) (NOT INFERRED) (Neither inductive nor deductive)

Contrapositive: verse?—see # 581 for very similar, but condition is stated "until I drink it new with you in the kingdom of God"

989. (if one is) . . . as the younger, (then that one shall be) . . . greatest among you . . . (Disc) (MA/GR-MR) (INFERRED) (Inductive) NOTE: INVERSION OF Original ORDERING
Contrapositive: verseF?—see # 260 for same conclusion, but condition is "servant", not "younger" – NOT in Matthew or Mark

990. (if one is) . . . as he that doth serve, (then that one shall be) . . . chief among you . . . (Disc) (MA/GR-MR) (INFERRED) (Inductive) NOTE: INVERSION OF Original ORDERING
Contrapositive: verseF?—see # 225 and 512 for the same

991. (since) . . . I (Jesus) appoint you a kingdom, . . . (then) . . . ye (the disciples who have continued with me in my temptations) may eat and drink at my table in my kingdom, and sit on thrones judging the twelve tribes of Israel. (Disc) (GAMR) (INFERRED) (Neither inductive nor deductive)
Contrapositive: verseF?—see # 210 for essentially similar, though this version has "appoint you a kingdom" in the condition and "eat and drink at my table" in the conclusion

992. . . . . when thou (Peter) art converted, (then) . . . strengthen thy brethren. (Disc) (MA-GA/MR) (INFERRED) (Neither inductive nor deductive)
Contrapositive: verseF? –Peter could not strengthen his brethren but still be converted – NOT in Matthew or Mark

993. . . . before the cock shall not crow this day . . . thou (Peter) shalt thrice deny that thou knowest me (Jesus), . . . (Peter) (MAGR) (NOT INFERRED-but inverted order) (Neither inductive nor deductive)
Contrapositive: verseF?—see same in # 383

994. . . . (if one) . . . hath a purse, (then) . . . let him take it . . . (Disc) (MAMR) (INFERRED) (Deductive)
Contrapositive: verseF? – One can choose not to take something, but still have it.
NOT in Matthew or Mark

995. . . . (if one . . . hath) . . . scrip, (then . . . let him take it) . . . (Disc) (MAMR) (INFERRED) (Deductive)

Contrapositive: verseF? - One can choose not to take something, but still have it.
NOT in Matthew or Mark

996. . . . (if one) . . . hath no sword, (then) . . . let him sell his garment, and buy one (a sword). (Disc) (MAMR) (INFERRED) (Deductive)
Contrapositive: verseF? - One can choose not to buy something, but still have it.
NOT in Matthew or Mark – Note that this does not state explicitly to use it.

997. (since) . . . I (Jesus) say unto you, that this that is written must yet be accomplished in me, And he was reckoned among the transgressors: for the things concerning me have an end . . . But now, he that hath a purse, let him take it, and likewise his scrip, and he that hath no sword, let him sell his garment, and buy one . . . (Disc) (GA-MA/MR) (INFERRED) (Neither inductive nor deductive)
Contrapositive: verse? – NOT in Matthew or Mark

998. . . . (since) . . . the things concerning me have an end, . . . he was reckoned among the transgressors . . . (Disc) (GA/GR-MR) (INFERRED) (Neither inductive nor deductive)
Contrapositive: verse? – NOT in Matthew or Mark

999. . . . .if thou (God the Father) be willing (to remove this cup of agony), (then) . . . (God the Father) remove this cup from . . . (Jesus Christ) (GAGR) (INFERRED – 'if' is given) (Neither inductive nor deductive)
Contrapositive: verse? – NOT in Matthew or Mark

1000. When I was daily with you in the temple, (then) . . . ye stretched forth no hands against me . . . (Relig) (GAMR) (INFERRED) (Neither inductive nor deductive)
Contrapositive: verseF?—see # 391 for same

1001. . . . If I tell you (the Jewish religious leaders) . . . (that I am the Christ), (then) ye will not believe . . . (Relig) (GAMR) (INFERRED- though "if" given) (Neither inductive nor deductive)
Contrapositive: verseF? – NOT in Matthew or Mark

1002. . . . if I (Jesus) also ask you, (then) ye will not answer me, nor let me go. (Relig) (GAMR) (INFERRED – though "if" given) (Neither inductive nor deductive)
Contrapositive: verseF? – NOT in Matthew or Mark

1003. . . . Hereafter (after this time of crucifixion and death), (then) . . . shall the Son of man sit on the right hand of the power of God. (Relig) (GAGR) (INFERRED) (Deductive)
Contrapositive: verseF? – NOT in Matthew or Mark

1004. (since) . . . behold the days are coming, in the which they shall say, Blessed are the barren, and the wombs that never bare, and the paps which never gave suck . . . Daughters of Jerusalem, weep not for me, but weep for yourselves, and for your children . . . (Company of women) (GA-MA/MR) (INFERRED) (Neither inductive nor deductive)
Contrapositive: verseF? – NOT in Matthew or Mark

1005. (when) . . . the days (have come) . . . , in which they shall say, Blessed are the barren, and the wombs that never bare, and the paps which never gave suck . . . Then shall they begin to say to the mountains, Fall on us; and to the hills, Cover us. (Company of women) (GA-MA/MR) (INFERRED- though "then" is given) (Deductive)
Contrapositive: verseF? – NOT in Matthew or Mark

1006. (since) .. they know not what they do . . . (then) . . . Father, forgive them (all who crucified Him) . . . (Mult) (MAGR) (INFERRED) (Neither inductive nor deductive)
Contrapositive: verseF? – NOT in Matthew or Mark

1007. (since) . . . a spirit hath not flesh and bones, as ye see me have . . . . Behold my hands and my feet that it is myself: handle me, and see . . . . (Disc) (GA-MA/MR) (INFERRED) (Neither inductive nor deductive)
Contrapositive: verseF? – NOT in Matthew or Mark

1008. (if you) . . . Behold my hands and my feet (then you will know) that it is myself . . . (Disc) (MA/MR-GR) (INFERRED) (Neither inductive nor deductive)
Contrapositive: verseF? – NOT in Matthew or Mark

1009. (if you) . . . handle me (then you will) . . . see (that it is myself) . . . (Disc) (MA/MR-GR) (INFERRED) (Neither inductive nor deductive)
Contrapositive: verseF? – Not in Matthew or Mark

1010. . . . while I was yet with you . . . These are the words which I spake unto you . . . that all must be fulfilled, which were written in the law of Moses,

and in the prophets, and in the psalms, concerning me. (Disc) (GAGR) (INFERRED) (Neither inductive nor deductive)
Contrapositive: verseF? – NOT in Matthew or Mark

1011. (since) . . . These are the words which I spake unto you . . . that all must be fulfilled, which were written in the law of Moses, and in the prophets, and in the psalms, concerning me . . . Thus it written, and thus it behoved Christ to suffer, and to rise from the dead the third day: And that repentance and remission of sins should be preached in his name among all nations, beginning at Jerusalem. (Disc) (GAGR) (INFERRED) (Deductive)
Contrapositive: verse? – NOT in Matthew or Mark

1012. . . . until ye be endued with power from on high . . . tarry ye in the city of Jerusalem . . . (Disc) (GAMR) (INFERRED) (Neither inductive nor deductive)
Contrapositive: verse? – NOT in Matthew or Mark

CONDITIONAL/PROCEDURAL FORMS IN THE GOSPEL OF JOHN

1013. Before that Philip called thee, when thou (Nathanael) wast under the fig tree, I (Jesus) saw thee. (Nathanael) (MAGR) (NOT INFERRED) (Neither inductive nor deductive) (Procedural – timing)
Contrapositive: verseF? – NOT in Matthew, Mark, or Luke

1014. Hereafter ye shall see heaven open, and the angels of God ascending and descending upon the Son of man. (Nathanael) (GA/MR-GR) (NOT INFERRED) (Neither inductive nor deductive)
Contrapositive: verseF? – NOT in Matthew, Mark, or Luke

1015. (If you) . . . Destroy this temple . . . (then) in three days I (Jesus) will raise it up. (Mult) (MAGR) (INFERRED) (Neither inductive nor deductive)
Contrapositive: verse? – NOT in Matthew, Mark, or Luke

1016. (if) . . . a man be (not) . . . born again, (then) . . . he cannot see the kingdom of God. (Nicodemus) (MA-GA/MR) (INFERRED) (Deductive)
Contrapositive: verse? – NOT in Matthew, Mark, or Luke

1017. (if) . . . a man be (not) . . . born of water and of the Spirit, (then) . . . he cannot enter into the kingdom of God. (Nicodemus) (MA-GA/MR) (INFERRED) (Deductive)
Contrapositive: verse? – NOT in Matthew, Mark, or Luke

1018. (if a thing) . . . is born of flesh, (then that thing) . . . is flesh. (Nicodemus) (GAMR) (INFERRED) (Deductive)
Contrapositive: verse? – NOT in Matthew, Mark, or Luke

1019. (if a thing) . . . is born of the Spirit, (then that thing) . . . is spirit. (Nicodemus) (GAMR) (INFERRED) (Deductive)
Contrapositive: verse? – NOT in Matthew, Mark, or Luke

1020. (if) . . . one . . . is born of the Spirt, (then that one is as) . . . the wind (that) . . . bloweth where it listeth, and thou hearest the sound thereof, but canst not tell whence it cometh, and whither it goeth . . . (Nicodemus) (GA/MR-GR) (Anal- symbolic only) (INFERRED) (Deductive)
Contrapositive: verse? – NOT in Matthew, Mark, or Luke – The converse also holds since this is a definition

1021. If I (Jesus) have told you earthly things, (then) . . . ye believe not . . . (Nicodemus) (GAMR) (INFERRED – "if" given) (Deductive)
Contrapositive: verse? – NOT in Matthew, Mark, or Luke (though see # 589 for same conclusion implying different condition and # 475 for opposite of conclusion implying yet another condition

1022. (if one is a) . . . man (then that one) hath (not) ascended up to heaven . . . (Nicodemus) (GAMR) (INFERRED) (Deductive)
Contrapositive: verseF? – NOT in Matthew, Mark, or Luke

1023. (if one) . . . believeth in him (the Son of man - Jesus), (then that one) . . . should not perish, but have eternal life (Nicodemus) (GA-MA/MR-GR) (INFERRED) (Deductive)
Contrapositive: verse? – NOT in Matthew, Mark, or Luke (SEE # 1127 for close to converse)

1024. (if a man does not) . . . perish, (then that man will) . . . have eternal life. (Nicodemus) (GA-MA/GR-MR) (INFERRED) (Inductive)
Contrapositive: verse? – NOT in Matthew, Mark, or Luke

1025. (since) . . . God so loved the world (then) . . . he (God) gave his only begotten Son . . . (Nicodemus) (GAGR) (INFERRED) (Deductive)
Contrapositive: verse? – NOT in Matthew, Mark, or Luke

1026. (since) . . . God so loved the world that he (God) gave his only begotten Son, (then) whosoever believeth in him (Jesus) should not perish, but have everlasting life. (Nicodemus) (GAMR) (INFERRED) (Deductive)
Contrapositive: verse? – NOT in Matthew, Mark, or Luke

1027. Repeat of #1023

1028. (since) . . . God sent not Son into the world to condemn the world; but that the world through him might be saved . . . . whosoever believeth in him should not perish, but have everlasting life. (Nicodemus) (GAMR) (INFERRED) (Deductive)
Contrapositive: verse? – NOT in Matthew, Mark, or Luke

{
1029. (if one) . . . believeth on him (Jesus), (then that one) is not condemned . . . (Nicodemus) (MA/GR-MR) (INFERRED) (Deductive)
Contrapositive: verse? – NOT in Matthew, Mark, or Luke (see #1023 FOR SAME CONDITION, but different conclusion with similar meaning)

1030. (if one) . . . believeth not (on him), (then that one) is condemned already . . . (Nicodemus) (MA/GR-MR) (INFERRED) (Deductive)
Contrapositive: verse? – NOT in Matthew, Mark, or Luke
} STINV

1031. . . . (since) . . . he hath not believed in the name of the only begotten Son of God . . . (then he) . . . is already condemned . . . (Nicodemus) (MA/MR-GR) (INFERRED) (Deductive)
Contrapositive: verse? – NOT in Matthew, Mark, or Luke

1032. . . . (since) . . . their deeds (men's) were evil . . . men loved darkness rather than light . . .
Contrapositive: verse? – NOT in Matthew, Mark, or Luke—see # 649 and # 650 for "contrapositive-like" and "converse-like", though not the same since "evil" is not the same "loved darkness rather than light"

1033. (if one) . . . doeth the truth, (than that one) . . . cometh to the light (Nicodemus) (Anal-s) (MA/MR-GR) (INFERRED) (Deductive)
Contrapositive: verseF? – NOT in Matthew, Mark, or Luke—Possible to obey the law, but not be in Jesus—- yet doeth the truth seems to imply that the person doeth what Jesus says as He is Truth

1034. (if one) . . . cometh to the light, (then that one's) . . . deeds may be made manifest that they are wrought in God. (Nicodemus) (Anal-s) (GA-MA/MR-GR) (INFERRED) (Deductive)
Contrapositive: verse? – NOT in Matthew, Mark, or Luke

1035. If thou knewest the gift of God, and who it is that saith to thee, Give me to drink; (then) . . . thou wouldest have asked of him, and he would have given thee living water. (Samaritan woman at well) (MA/MR-GR) (INFERRED – though "if" given) (Deductive)
Contrapositive: verseF? – we can not be asking God for things, but still know things about God – NOT in Matthew, Mark, or Luke

1036. (if one) . . . drinketh of this water (physical water) , (then that one) . . . shall thirst again . . . (Samaritan woman at the well) (MAMR) (INFERRED) (Inductive)
Contrapositive: verseF? – NOT in Matthew, Mark, or Luke

1037. (if one) . . . drinketh of the water that I shall give him (spiritual) , (then that one) . . . shall never thirst again . . . (Samaritan woman at the well) (Anal – s) (MA-GA/MR) (INFERRED) (Deductive)
Contrapositive: verse? – NOT in Matthew, Mark, or Luke—see form # 1037 for the conclusion of # 1118, but the negation of the condition of # 1118.

1038. (since) . . . thou (the Samaritan woman at the well) hast had five husbands; and he whom thou now hast is not thy husband . . . , Thou hast well said, I have no husband . . . (Samaritan woman at the well) (MAMR) (INFERRED) (Neither inductive nor deductive)
Contrapositive: verseF? – NOT in Matthew, Mark, or Luke

1039. (since) . . . salvation is of the Jews (then) . . . we (the Jews) know what we worship . . . (Samaritan woman at the well) (GAMR) (INFERRED) (Deductive)
Contrapositive: verseF? – NOT in Matthew, Mark, or Luke

1040. (since) . . . the Father seeketh such to worship him (then) . . . the hour cometh, and now is, when the true worshippers shall worship the Father in spirit and truth . . . (Samaritan woman at the well) (GA/GR-MR) (INFERRED) (Deductive)
Contrapositive: verseF? – NOT in Matthew, Mark, or Luke

1041. (If one) . . . worship(s) him (God) (then that one) must worship him in spirit and in truth. (Samaritan woman at the well) (MAMR) (INFERRED) (Deductive)
Contrapositive: verse? – NOT in Matthew, Mark, or Luke

1042. (since) . . . they (the fields) are white already unto harvest (then) . . . Lift up your eyes, and look on the fields . . . (Samaritan woman at the well) (Anal-literal) (GAMR-GAMR) (INFERRED) (Deductive)
Contrapositive: verseF? – NOT in Matthew, Mark, or Luke

1043. (if one) .. reapeth (then that one) . . . receiveth wages, and gathereth fruit unto eternal life . . . (Samaritan woman at the well) (MA-GA/MR) (Anal-s) (INFERRED) (Deductive)
Contrapositive: verse? – NOT in Matthew, Mark, or Luke

1044. (if one) . . . reapeth (and) receiveth wages, and gathereth unto life eternal, (then) both he that soweth and he that reapeth may rejoice together. (Samaritan woman at the well) (MAGA-MR) (Anal-s) (INFERRED) (Deductive)
Contrapositive: verseF? – NOT in Matthew, Mark, or Luke

1045. (if you do not) . . . see signs and wonders, (then) . . . ye will not believe. (Nobleman at Capernaum) (MAMR) (INFERRED) (Inductive)—seems to apply to all unbelievers, but stated only to this man)
Contrapositive: verseF? - NOT in Matthew, Mark, Luke - KEY EXCEPTION NOTED for formal human human reasoning—see # 1396; also # 304 seems to show that people who believe will see signs and wonders, though by false Christs

1046. (if one) . . . sin(s) more, (then) . . . a worse thing (than being impotent) (will) come unto thee. (Healed impotent man) (MA-GR/MR) (INFERRED) (Inductive)
Contrapositive: verse? – NOT in Matthew, Mark, or Luke

1047. (if) . . . he (the Father) doeth (a thing), (then) . . . these also doeth the Son likewise . . . (Relig) (GAGR) (INFERRED) (Neither inductive nor deductive, but specific to God and Christ)
Contrapositive: verse? – NOT in Matthew, Mark, or Luke

1048. (since) . . . what things soever he (the Father) doeth, these also doeth the Son likewise . . . (then) The Son can do nothing of himself, but what he seeth the Father do . . . (Relig) (GAGR) (INFERRED) (Neither inductive nor deductive, but specific to God and Christ)
Contrapositive: verse? - NOT in Matthew, Mark, or Luke

1049. (since) . . . the Father loveth the Son, and sheweth him all things that himself doeth, (then) . . . what things soever he (the Father) doeth, these also

doeth the Son likewise . . . (Relig) (GAGR) (INFERRED) (Neither inductive nor deductive, but specific to God and Christ)
Contrapositive: verse? – NOT in Matthew, Mark, or Luke

1050. (When) . . . he (the Father) will shew him (the Son) greater works than these, (then) you (the religious leaders) may marvel. (Relig) (GAMR) (INFERRED) (Neither inductive nor deductive)
Contrapositive: verseF? – NOT in Matthew, Mark, or Luke

1051. (since) . . . as the Father raiseth up the dead, and quickeneth them, (then) . . . he (the Father) will shew him (the Son) greater works than these, . . . (Relig) (GAGR) (INFERRED) (Deductive)
Contrapositive: verse? – NOT in Matthew, Mark, or Luke

1052. (since) . . . the Father raiseth up the dead, and quickeneth them, (then) . . . even so the Son quickenth whom he will (Relig) (GAGR) (INFERRED) (Deductive)
Contrapositive: verse? – NOT in Matthew, Mark, or Luke

1053. (since) . . . the Father judgeth no man, but hath committed all judgment unto the Son, (then) . . . the Son quickeneth whom he will. (Relig) (GAGR) (INFERRED) (Deductive)
Contrapositive: verse? – NOT in Matthew, Mark, or Luke

1054. (since) . . . the Father judgeth no man, but hath committed all judgment unto the Son, (then) . . . all men should honor the Son, even as they honor the Father . . . (Relig) (GAMR) (INFERRED) (Deductive)
Contrapositive: verse? – vaccuously – NOT in Matthew, Mark, or Luke

1055. (if one) . . . honoureth not the Son (then that one) . . . honoureth not the Father which hath sent him. (Relig) (MAMR) (INFERRED) (Deductive)
Contrapositive: verse? – NOT in Matthew, Mark, or Luke

1056. (if one) . . . heareth my word, and believeth on him that sent me, (then that one) . . . hath everlasting life, and shall not come into condemnation, but is passed from death unto life. (Relig) (MA/MR-GR) (INFERRED) (Inductive)
Contrapositive: verseF? – NOT in Matthew, Mark, or Luke—see # 1023 and # 1029 for interesting forms related—see #1215 for hear and not believe, with opposite conclusion to condemnation (i.e., not judged)

1057. (since) . . . as the Father hath life in Himself, (then) . . . The hour is coming, and now is, when the dead shall hear the voice of the Son of God: and they that hear shall live . . . (Relig) (GA/MR-GR) (INFERRED) (Deductive)
Contrapositive: verseF? – NOT in Matthew, Mark, or Luke

1058. (if one who is dead) . . . hears(s the voice of God then that one) . . . shall live. (Relig) (GA/MR-GR) (INFERRED) (Inductive)
Contrapositive: verseF? – NOT in Matthew, Mark, or Luke

1059. (since) . . . the Father hath life in himself (then) . . . so hath he given to the Son to have life in himself; And hath given him authority to execute judgment also . . . (Relig) (GAGR) (INFERRED) (Neither inductive nor Deductive)
Contrapositive: verse? – NOT in Matthew, Mark, or Luke

1060. (since) . . . he (Jesus) is the Son of man, (then) . . . (the Father) . . . hath . . . given to the Son to have life in himself; And hath given him authority to execute judgment also . . . (Relig) (GAGR) (INFERRED) (Neither inductive nor Deductive)
Contrapositive: verse? – verse? – NOT in Matthew, Mark, or Luke

1061. (since) . . . the hour is coming, in the which all that are in the graves shall hear his voice, And shall come forth . . . (then) . . . (ye the religious) Marvel not at this (that the Son hath the Father's life in himself and the authority to execute judgment) . . . (Relig) (GAMR) (INFERRED) (Neither inductive not deductive)
Contrapositive: verseF? – If men don't marvel, then hour is still coming—NOT in Matthew, Mark, or Luke

1062. (if) they (those who are in the graves) . . . have done good (then they will have) . . . the resurrection of life . . . (Relig) (MA/GR-MR) (INFERRED) (Deductive)
Contrapositive: verse? – NOT in Matthew, Mark, or Luke (though there are a couple of passages that have same basic meaning in other words)

1063. (if) they (those who are in the graves) . . . have done evil (then they will have) . . . the resurrection of damnation . . . (Relig) (MA/GR-MR) (INFERRED) (Deductive)
Contrapositive: verse? - NOT in Matthew, Mark, or Luke (though there is one other passage that has the same basic meaning in other words)

1064. (whenever) . . . I (Jesus) hear (then) . . . I judge; and my judgment is just . . . (Relig) (GAGR) (INFERRED) (Neither inductive nor deductive)
Contrapositive: verse? – NOT in Matthew, Mark, or Luke

1065. (since) . . . I seek not mine own will, but the will of the Father which hath sent me, (then) . . . my judgment is just . . . (Relig) (GAGR) (INFERRED) (Neither inductive nor deductive)
Contrapositive: verse? Vacuously – NOT in Matthew, Mark, or Luke

1066. If I (Jesus) bear witness of myself, (then) . . . my witness is not true. (Relig) (GAGR) (INFERRED- though "if" is given) (Neither inductive nor deductive)
Contrapositive: verse? – NOT in Matthew, Mark, or Luke

1067. (when) . . . I (Jesus) say these things (about Jesus' relationship to the Father and about John the Baptist) (then) . . . ye (religious & Jews) might be saved. (Relig) (GAMR) (INFERRED) (Deductive)
Contrapositive: verse? – NOT in Matthew, Mark, or Luke

1068. (since) . . . the works which the Father hath given me to finish, the same works that I do, bear witness of me, that the Father hath sent me (then) . . . I have greater witness than that of John . . . (Relig) (GAGR) (INFERRED) (Neither inductive nor deductive)
Contrapositive: verse? – NOT in Matthew, Mark, or Luke

1069. (since) . . . whom he (the Father) hath sent, him ye believe not (then) . . . Ye have neither heard his voice at any time nor seen his shape and ye have not his word abiding in you . . . (Relig) (MAMR) (INFERRED) (Neither inductive not deductive)
Contrapositive: verse? – NOT in Matthew, Mark, or Luke

1070. (since) . . . in them (the scriptures) ye think ye have eternal life: and they (the scriptures) are they which testify of me . . . (then) Search the scriptures . . . (Relig) (MA-GA/MR) (INFERRED) (Deductive)
Contrapositive: verse? – NOT in Matthew, Mark, or Luke

1071. (if you) . . . . come unto me, (then) ye might have life. (Relig) (MA/GR-MR) (INFERRED) (Deductive)
Contrapositive: verse? – NOT in Matthew, Mark, or Luke

1072. if another shall come in his own name, (then) . . . him ye (Religious – Jews) will receive. (Relig) (MAMR) (INFERRED) (Inductive)
Contrapositive: verseF? – what one man does or does not do will not necessarily affect another man's actions – NOT in Matthew, Mark, or Luke

1073. (since) . . . had ye believed Moses . . . (then) there is one that accuseth you, even Moses . . . (Relig) (MAMR) (INFERRED) (Deductive)
Contrapositive: verseF? – NOT in Matthew, Mark, or Luke

1074. (if) . . . had ye believed Moses (then) . . . ye would have believed me (Relig) (MAMR) (INFERRED) (Deductive)
Contrapositive: verseF? – NOT in Matthew, Mark, or Luke (see # 881 for close to STINV)

1075. (since) he (Moses) wrote of me . . . (then) ye would have believed in me . . . (Relig) (MAMR) (INFERRED) (Deductive)
Contrapositive: verseF? – NOT in Matthew, Mark, or Luke

1076. (If you – the disciples) gather up the fragments that remain, (then) . . . nothing (will) be lost. (Disc) (MAMR) (INFERRED) (Deductive)
Contrapositive: verse? – NOT in Matthew, Mark, or Luke

1077. (since) . . . It is I (Jesus – who is here with you then) . . . be not afraid. (Disc) (GAMR) (INFERRED) (Deductive)
Contrapositive: verseF?—see # 453 for same

1078. (since) . . . ye (Mult) did eat of the loaves, and were filled (not because of the miracles), ... Ye (Mult) seek me . . . (Mult) (MAMR) (INFERRED) (Deductive)
Contrapositive: verseF? – NOT in Matthew, Mark, or Luke

1079. (since) . . . him (the Son of man) hath God the Father sealed . . . (then) Labour not for the meat which perisheth, but for that meat which endureth unto everlasting life, which the Son of man shall give unto you . . . (Mult) (GAMR) (INFERRED) (Deductive)
Contrapositive: verseF? – NOT in Matthew, Mark, Luke

1080. (if a thing is) . . . the work of God, (then it is) . . . believ(ing) on him whom he hath sent. (Mult) (GA/MR-GR) (INFERRED) (Deductive)

Contrapositive: verse? – NOT in Matthew, Mark, or Luke (NOTE that converse if not explicitly provided as would be the case with a definition in strict rules of human reasoning)

1081. (since) . . . the bread of God is he which cometh down from heaven, and giveth life unto the world, (then) . . . Moses gave you not that bread from heaven; but my Father giveth you the true bread from heaven . . . (Mult) (GA/MR-GR) (INFERRED) (Deductive)
Contrapositive: verseF? – NOT in Matthew, Mark, or Luke

1082. (if one) . . . cometh to me, (then that one) . . . shall never hunger . . . (Mult) (MAGR) (INFERRED) (Inductive) (Anal-s; condition is literal, but conclusion is symbolic)
Contrapositive: verse? – NOT in Matthew, Mark, or Luke

1083. (if one) . . . believeth on me, (then that one) . . . shall never thirst . . . (Mult) (MAGR) (INFERRED) (Inductive) (Anal-s; condition is literal, but conclusion is symbolic)
Contrapositive: verse? – NOT in Matthew, Mark, or Luke—see form # 1037 for the conclusion of # 1118, but the negation of the condition of # 1118.

1084. (if a thing is such that) . . . the Father giveth (it) me (Jesus), (then that thing) . . . shall come to me . . . (Mult) (GAGR) (INFERRED) (Deductive)
Contrapositive: verse? – NOT in Matthew, Mark, Luke

1085. (if one) cometh to me, (then) I will in no wise cast out (him) . . . (Mult) (MAGR) (INFERRED) (Inductive)
Contrapositive: verse? – NOT in Matthew, Mark, or Luke

1086. (since) . . . I (Jesus) came down from heaven, not to do mine own will, but the will of him that sent me . . . (then) him that cometh to me (Jesus) I will in no wise cast out . . . (Mult) (GAGR) (INFERRED) (Deductive)
Contrapositive: verse? – NOT in Matthew, Mark, or Luke

1087. (If a thing) is . . . the Father's will which that sent me (Jesus), (then that thing is such that) . . . of all which he hath given me I should lose nothing, but should raise up again at the last day. (Mult) (GAGR) (INFERRED) (Deductive)
Contrapositive: verse? – NOT in Matthew, Mark, or Luke (see John 17: 12 for instance where one is lost, but not by Jesus); the converse holds since this is a definition.

1088. (if) . . . he (the Father) hath given me (a thing), (then) I (Jesus) should lose nothing . . . (Mult) GAGR) (INFERRED) (Deductive)
Contrapositive: verse? – NOT in Matthew, Mark, or Luke

1089. (If a thing) is the will of him that sent me, (then) . . . every one which seeth the Son, and believeth on him, may have everlasting life: and I will raise him up at the last day. (Mult) (GA/MR-GR) (INFERRED) (Deductive)
Contrapositive: verse? – NOT in Matthew, Mark, or Luke – The converse holds since this is a definition, but notably the conclusion of # 1087 is not connected in any of the statements attributed to Christ in the Holy Bible to the conclusion of # 1089 so that no complete "transitivity" can be assumed here, though both do interestingly have in common that (Christ) will "raise up . . . at the last day" both those whom (God the Father) "hath given to (Christ)" and also to those who ". . . seeth him and believeth on him . . .".

1090. (if one) . . . seeth the Son, and believeth on him, (then that one) may have everlasting life: and I will raise him up at the last day. (Mult) (GA-MA/MR-GR) (INFERRED) (Deductive)
Contrapositive: NOT in Matthew, Mark, or Luke

1091. (if) . . . the Father which hath sent me (does not) draw him, . . . (then that man) can(not) come to me . . . (Relig) (GAMR) (INFERRED) (Deductive)
Contrapositive: verse? – NOT in Matthew, Mark, or Luke

1092. (if one is a child of the Gentiles), (then that one) . . . shall be taught of God. (Relig) (GAGR) (INFERRED) (Deductive)
Contrapositive: verse? – NOT in Matthew, Mark, or Luke, but see Isaiah 54:13 for context to provide condition

1093. (since all of the children of the Gentiles) . . . shall be all taught of God, Every man therefore that hath heard, and hath learned of my Father, cometh unto me. (Relig) (GA/GR-MR) (INFERRED- though therefore given) (Inductive)
Contrapositive: verse? – NOT in Matthew, Mark, or Luke—see # 1125 for close to converse (though "known" and "cometh to" are different)

1094. (if a) . . . man . . . hath heard, and hath learned of the Father, (then that one) . . . cometh unto me. (Relig) (MA-GA/MR) (INFERRED) (Deductive)
Contrapositive: verse? – NOT in Matthew, Mark, or Luke (see # 1094 for somewhat like STINV, but "given" and "learned" are different and additional

condition to # 109 ("hath heard")—see # 1055 for somewhat like a contrapositive to # 1055, though "hath head and hath learned" are different than "honoureth"

1095. (if one is a man), (then that one) hath not seen the Father, save he which is of God, he hath seen the Father. (Relig (GA/MR-GR) (INFERRED) (Deductive)
Contrapositive: verse? – NOT in Matthew, Mark, or Luke—see # 1250 for counterexample in which disciples "from henceforth" have seen the Father (same Greek word for "seen" since this statement is not qualified in time

1096. See same in # 1023 (though this form does not have "perish not"

1097. (If a thing) . . . is the bread (the bread of life) which cometh down from heaven (then that thing is such) . . . that a man may eat thereof, and not die. (Relig) (Anal-s) (GAMR) (INFERRED) (Deductive)
Contrapositive: verseF? – NOT in Matthew, Mark, or Luke – The converse holds since this is a definition.

1098. (if) . . . a man eat(s) thereof (of the bread which cometh down from heaven), (then that man will) . . . not die. (Relig) (Anal-s) (MA-GA/MR) (INFERRED) (Inductive)
Contrapositive: verse? – NOT in Matthew, Mark, or Luke

{
1099. if any man eat of this bread (living bread which came down from heaven), (then) he shall live forever . . . (Relig) (Anal-s) (MAGR) (INFERRED) (Inductive)
Contrapositive: verse? – NOT in Matthew, Mark, or Luke (see # 1107 for same)

{
1100. (if any one does not) . . . eat the flesh of the Son of man, and drink his blood, (then that one) . . . has no life . . . (Relig) (Anal-s) (MAGR) (INFERRED) (Deductive)
Contrapositive: verse? – NOT in Matthew, Mark, or Luke
} STINV somewhat close to 1099

1101. (if any one) . . . eateth the flesh of the Son of man, and drinketh his blood, (then that one) . . . hath eternal life and I will raise him up at the last day . . . (Relig) (Anal-s) (MAGR) (INFERRED) (Inductive)

Contrapositive: verse? – NOT in Matthew, Mark, or Luke
} STINV close to 1100 (though extra part to conclusion in 1101)

1102. (since) . . . my flesh is meat indeed, and my blood is drink indeed . . . . (then) . . . any one) . . . eateth the flesh of the Son of man, and drinketh his blood, . . . hath eternal life and I will raise him up at the last day . . . (Relig) (Anal-s) (GA/MR-GR) (INFERRED) (Inductive)
Contrapositive: verse? – NOT in Matthew, Mark, or Luke

1103. (if any one) . . . eateth the flesh of the Son of man, and drinketh his blood, (then that one) . . . dwelleth in me, and I in him. (Relig) (Anal-s) (MAGR) (INFERRED) (Deductive)
Contrapositive: verse? – NOT in Matthew, Mark, or Luke

1104. As the living Father hath sent me, and I live by the Father: so he that eateth me, even he shall live by me. (Relig) (Anal-s) (GAMR) (NOT INFERRED) (Inductive)
Contrapositive: verse? – NOT in Matthew, Mark, or Luke (NOTE: only part of conclusion is "symbolic" – "eateth me"; the rest of condition and conclusion are literal)

1105. (if one) . . . eateth me, (then that one) . . . shall live by me. (Relig) (Anal-s) (MA/MR-GR) (INFERRED) (Inductive)
Contrapositive: verse? – NOT in Matthew, Mark, or Luke (NOTE: Condition is "symbolic", but conclusion is real)

1106. (if one) . . . eateth of this bread . . . that came down from heaven: not as your fathers did eat manna, (then that one) . . . shall live for ever. (Relig) (MA-GA/MR-GR) (INFERRED) (Inductive)
Contrapositive: verse? – See # 1099 for same

1107. (if a) . . . word (is such that) . . . I (Jesus) speak (it) unto you, (then that word(s)) . . . (is) are spirit and . . . (is) are life. (Disc) (GAGR) (INFERRED) (Deductive)
Contrapositive: verse? – NOT in Matthew, Mark, or Luke

1108. (if) . . . it were not given unto him of my Father, (then) . . . no man can come unto me . . . (Disc) (GAMR) (INFERRED) (Deductive)
Contrapositive: verse? – NOT in Matthew, Mark, or Luke {see # 1094 for somewhat like STINV, but "given" and "learned" are different and additional condition to # 109 ("hath heard")}

1109. . . . (since) I testify of it (the world), that the works thereof are evil . . . (then) me (Jesus) it hateth . . . (Jesus's brethren) (GAMR) (INFERRED) (Deductive)
Contrapositive: verseF? – NOT in Matthew, Mark, or Luke

1110. (since) . . . my time is not yet full come . . . (then) I go not up yet unto this feast . . . (Jesus' brethren) (GAGR) (INFERRED) (Deductive)
Contrapositive: verse? – NOT in Matthew, Mark, or Luke

1111. If any man will do his will (the Father's will), (then) . . . he shall know of the doctrine (that Jesus taught) whether it be of God, or whether I speak of myself. (Relig) (MA/MR-GR) (INFERRED- though "if" given) (Inductive)
Contrapositive: verse? – NOT in Matthew, Mark, or Luke

1112. (if one) . . . speaketh of himself, (then that one) . . . seeketh his own glory . . . (Relig) (MAMR) (INFERRED) (Deductive)
Contrapositive: verseF? – NOT in Matthew, Mark, or Luke

1113. (if one) . . . seeketh his glory that sent him, (then that one) . . . is true, and no unrighteousness is in him . . . (Relig) (MAMR) (INFERRED) (Deductive)
Contrapositive: verse? – NOT in Matthew, Mark, or Luke

1114. (if one is a Jew (Pharisee, scribe), then that one does) . . . not . . . keepeth the law. (Relig) (GAMR) (INFERRED) (Deductive)
Contrapositive: verse? – vaccuously, since no one keepeth the law – NOT in Matthew, Mark, or Luke

1115. (since) . . . I (Jesus) have done one work, and ye all marvel. Moses therefore gave unto you circumcision; (not because it is of Moses, but of the fathers;) . . . (Relig) (GAMR) (INFERRED – though "therefore" is given) (Deductive)
Contrapositive: verseF? – the two seem completely disconnected in human reasoning – NOT in Matthew, Mark, or Luke

1116. If a man on the Sabbath day receive circumcision, (then) that the law of Moses should not be broken . . . (Relig) (MAMR) (NOT INFERRED) (Deductive)
Contrapositive: verseF? – NOT in Matthew, Mark, or Luke

1117. (since) . . . I am from him, and he hath sent me (the Father), (then) . . . I (Jesus) know him (the Father) . . . (Mult) (GAGR) (INFERRED) (Neither inductive nor deductive)
Contrapositive: verse? – NOT in Matthew, Mark, or Luke

1118. If any man thirst, (then) . . . let him come unto me, and drink. (Mult) (Anal –s) (MAMR) (INFERRED- though "if" is given) (Deductive)
Contrapositive: verseF?—see form # 1037 for the conclusion of # 1118, but the negation of the condition of # 1118.

1119. (if one) . . . believeth on me, (then) . . . as the scripture hath said, out of his belly shall flow rivers of living water. (Mult) (Anal-s) (MA-GA/MR-GR) (INFERRED) (Inductive)
Contrapositive: verse? – NOT in Matthew, Mark, or Luke

1120. (if one) . . . is without sin among you (the accusers of the woman taken in adultery), (then) . . . let him first cast a stone at her. (Relig) (MAMR) (INFERRED) (Deductive)
Contrapositive: verseF? – NOT in Matthew, Mark, or Luke

1121. (if one) . . . followeth me, (then that one) . . . shall not walk in darkness, but shall have the light of life. (Relig) (MA/MR-GR) (INFERRED) (Inductive)
Contrapositive: verse? – NOT IN Matthew, Mark, or Luke—see # 1204 for close to converse

1122. (since) . . . I (Jesus) know whence I came, and whither I go, (then) . . . though I bear record of myself, yet my record is true . . . (Relig) (GAGR) (INFERRED) (Neither inductive nor deductive)
Contrapositive: verse? - NOT in Matthew, Mark, or Luke

1123. if I (Jesus) judge, (then) . . . my judgment is true . . . . (Relig) (GAGR) (INFERRED, though "if" given) (Neither inductive nor deductive)
Contrapositive: NOT in Matthew, Mark, or Luke

1124. (since) . . . I am not alone, but I and the Father that hath sent me, (then) . . . my judgment is true. (Relig) (GAGR) (INFERRED) (Neither inductive nor deductive)
Contrapositive: NOT in Matthew, Mark, or Luke

1125. . . . if ye had known me (Jesus), (then) ye should have known my Father also. (Relig) (MA-GA/MR-GR) (INFERRED) (Deductive)

Contrapositive: NOT in Matthew, Mark, or Luke—see # 1093 for close to converse (though "known" and "cometh to" are different)—see # 1249 for same as # 1125—see # 1055 for somewhat of an inverse, though "known" and honoureth" are different

1126. (since) . . . ye are from beneath; I am from above; ye are of this world; I am not of this world (, then that thing) I (Jesus) said therefore unto you (the religious) that ye shall die in your sins . . . (Relig) (MA-GA/GR-MR) (INFERRED) (Deductive)
Contrapositive: verse? – NOT in Matthew, Mark, or Luke

1127. if ye believe not that I (Jesus) am he, (then) ye shall die in your sins. (Relig) (MAMR) (INFERRED – though "if" given) (Inductive)
Contrapositive: verse? – NOT in Matthew, Mark, or Luke (see # 1023 for close to converse)

1128. (since) . . . if ye believe not that I (Jesus) am he, I (Jesus) said therefore unto you (the religious) that ye shall die in your sins . . . (Relig) (MA/GR-MR) (INFERRED) (Inductive)
Contrapositive: verseF? – NOT in Matthew, Mark, or Luke

1129. When ye have lifted up the Son of man, then shall ye know that I am he, and that I do nothing of myself, but as my Father hath taught me, I speak these things. (Relig) (MA/MR-GR) (NOT INFERRED) (Deductive)
Contrapositive: verse? – NOT in Matthew, Mark, or Luke

1130. . . . (since) . . . I (Jesus) do always those things that please him . . . . (then) the Father hath not left me alone . . . (Relig) (GAGR) (INFERRED) (Neither deductive nor inductive)
Contrapositive: verse? – NOT in Matthew, Mark, or Luke
(NOTE: See "My God, my God, why hast thou forsaken me?" for extra condition (Jesus taking on the sins of the world) to # 1130 that causes it not to be true, though this does not invalidate # 1130)

1131. If ye continue in my word, then are ye my disciples indeed; and ye shall know the truth, and the truth shall make you free. (Relig) (MA/MR-GR) (Anal-literal) (NOT INFERRED) (Inductive)
Contrapositive: verse? – NOT in Matthew, Mark, or Luke

1132. (if) . . . ye shall know the truth (I think Jesus is referring to himself and to His Word), (then) . . . the truth shall make you free. (Relig) (MA-GA/GR-MR) (Anal-literal) (INFERRED) (Deductive)
Contrapositive: verse? – NOT in Matthew, Mark, or Luke

1133. (if one) . . . committeth sin, (then that one) . . . is the servant of sin. (Relig) (MAMR) (INFERRED) (Deductive)
Contrapositive: verse? – NOT in Matthew, Mark, or Luke

1134. (if one is a) . . . servant, (then that one) . . . abideth not in the house for ever . . . (Relig) (MAMR) Anal-literal) (INFRERRED) (Deductive)
Contrapositive: verseF? – NOT in Matthew, Mark, or Luke

1135. If the Son therefore shall make you free, (then) . . . ye shall be free indeed. (Relig) (GAMR) (Anal-s) (INFERRED, though IF given) (Deductive)
Contrapositive: verse? – NOT in Matthew, Mark, or Luke

1136. (since) . . . my word (Jesus's word) hath no place in you, (then) . . . ye seek to kill me (Jesus) . . . (Relig) (GAMR) (INFERRED) (Deductive)
Contrapositive: verse? – NOT in Matthew, Mark, or Luke

1137. If ye were Abraham's children, (then) ye would do the works of Abraham. (Relig) (MAMR) (INFERRED- though IF given) (Deductive) (Anal-literal)
Contrapositive: verse? – NOT in Matthew, Mark, or Luke

1138. If God were your Father, (then) ye would love me . . . (Relig) (MAMR) (INFERRED – though "If" is given) (Deductive)
Contrapositive: verse? – NOT in Matthew, Mark, or Luke—see # 1055 for somewhat like contrapositive, though "love" and "honoureth" are different

1139. (since) . . . I (Jesus) proceeded forth and came from God; neither came I of myself, but he sent me (then) . . . If God were your Father, (then) ye would love me . . . (Relig) (MAMR) (INFERRED) (Deductive)
Contrapositive: verse? – NOT in Matthew, Mark, or Luke

1140. . . . because ye cannot hear my word, (then) ye (do) not understand my speech . . . (Relig) (MA-GA/MR-GR) (Anal – literal) (INFERRED – THOUGH "BECAUSE" IS GIVEN) (Deductive)
Contrapositive: verse? – NOT in Matthew, Mark, or Luke (see # 127 to show that inverse need not be true)

1141. (since) . . . there is no truth in him (the devil), (then) . . . He (the devil) was a murderer from the beginning, and abode not in the truth . . . (Relig) (Satan-Satan) (INFERRED) (Neither inductive nor deductive)
Contrapositive: verse? – NOT in Matthew, Mark, or Luke

1142. When he (the devil) speaketh a lie, (then) he speaketh of his own . . . (Relig) (Satan-Satan) (INFERRED- though "When" given) (Neither inductive nor deductive)
Contrapositive: verse? – NOT in Matthew, Mark, or Luke

1143. (since) . . . he is a liar, and the father of it, (then) he speaketh of his own . . . (Relig) (Satan-Satan) (INFERRED- though "When" given) (Neither inductive nor deductive)
Contrapositive: verse? – NOT in Matthew, Mark, or Luke

1144. (since) . . . I (Jesus) tell you the truth, (then) ye believe me not. (Relig) (GAMR) (INFERRED) (Deductive)
Contrapositive: verse? – NOT in Matthew, Mark, or Luke

{
1145. (if one) . . . is of God (then that one) . . . heareth God's Words . . . (Relig) (MA-GA/MR) (INFERRED) (Deductive)
Contrapositive: verse? – NOT in Matthew, Mark, or Luke

1146. (since) ye are not of God . . . ye therefore hear them (God's Words) not . . . (Relig) (MA-GA/MR) (INFERRED) (Deductive) Contrapositive: verse? – NOT in Matthew, Mark, or Luke
} STINV

1147. If a man keep my saying, (then) he shall never see death. (Relig) (Anal-literal) (MA/MR-GR) (INFERRED – though "IF" is given) (Inductive)
Contrapositive: verse? – NOT in Matthew, Mark, or Luke (See # 205 for close to converse)

1148. If I honour myself, (then) my honour is nothing . . . (Relig) (GAGR) (INFERRED- though "If" is given) (Neither inductive nor deductive)
Contrapositive: verse? – NOT in Matthew, Mark, or Luke

1149. . . . if I should say, I know him not, (then) I shall be a liar like unto you . . . (Relig) (GAGR) (INFERRED-though "If" is given) (Neither inductive nor deductive)
Contrapositive: verse? – NOT in Matthew, Mark, or Luke

1150. Before Abraham was, I (Jesus) am. (Relig) (MA-GA/GR) (NOT INFERRED) (Neither inductive nor deductive)
Contrapositive: verse? – NOT in Matthew, Mark, or Luke

1151. . . . while it is day, (then) . . . I must work the works of him that sent me . . . (Relig) (GAGR) (INFERRED – though "when" is given) (Neither inductive nor deductive)
Contrapositive: verse? – NOT in Matthew, Mark, or Luke

1152. . . . . (when) . . . the night cometh, (then) . . . no man can work. (Relig) (Anal-literal) (GAMR) (INFERRED) (Deductive)
Contrapositive: verseF? – NOT in Matthew, Mark, or Luke

1153. As long as I (Jesus) am in the world, (then) I am the light of the world. (Relig) (GAGR) (Anal-literal) (INFERRED- though "as" given) (Deductive)
Contrapositive: verse? – NOT in Matthew, Mark, or Luke

1154. (Since) judgment (is to be done then) I am come into this world . . . (Disc) (GAGR) (INFERRED) (Neither inductive nor deductive)
Contrapositive: verse? – NOT in Matthew, Mark, or Luke

1155. (when) I am come into this world (then) . . . they which see not might see; and that they which see might be made blind. (Disc) (GA/MR-GR) (Anal-literal) (INFERRED) (Deductive)
Contrapositive: NOT in Matthew, Mark, or Luke

1156. If ye (Relig) were blind, (then) ye should have no sin . . . (Relig) (MA/MR-GR) (Anal-s) (INFERRED- though "if" given) (Deductive)
Contrapositive: verse? – NOT in Matthew, Mark, or Luke

1157. (since) . . . ye (Relig) say, We see; therefore, your sin remaineth . . . (Relig) (MA/MR-GR) (Anal-s) (INFERRED- though "if" given) (Deductive)
Contrapositive: verseF? – NOT in Matthew, Mark, or Luke

{
1158. (if one) . . . entereth not by the door into the sheepfold, but climbeth up some other way, (then that one) . . . is a thief and a robber. (Relig) (Anal-literal) (MAMR-Satan/Satan) (INFERRED) (Deductive for literal)
Contrapositive: verse? – NOT in Matthew, Mark, or Luke

1159. (if one) . . . entereth in by the door (into the sheepfold), (then that one) . . . is the shepherd of the sheep. (Relig) (Anal-literal) (MAMR-Jesus/Jesus) (INFERRED) (Deductive for literal)
Contrapositive: verse? – NOT in Matthew, Mark, or Luke
} close to STINV

1160. (when the shepherd is come to the door then) . . . the porter openeth and the sheep hear his voice: and he calleth his own sheep by name, and leadeth them out. (Relig) (MAMR/animals-GAGR) (Anal-literal) (INFERRED) (Deductive)
Contrapositive: verse? – NOT in Matthew, Mark, or Luke

1161. when he putteth forth his own sheep, (then) he goeth before them, and the sheep follow him . . . (Relig) (MA/animals-GA/GR-MR) (Anal-literal) (INFERRED) (Deductive)
Contrapositive: verse? – NOT in Matthew, Mark, or Luke

1162. (since) . . . they (the sheep) know his voice . . . (then) the sheep follow him . . . (Relig) (Animals/Animals-MAMR) (Anal-literal) (INFERRED) (Deductive)
Contrapositive: verse? – NOT in Matthew, Mark, or Luke

1163. (If one is) . . . a stranger (then that one) they (the sheep) will not follow, but will flee from him . . . (Relig) (Anal-literal) (MA/animals-Satan/MR) (INFERRED) (Inductive)
Contrapositive: verse? – NOT in Matthew, Mark, or Luke

1164. . . . (since) they (the sheep) know not the voice of strangers . . . (then) they (the sheep) will not follow, but will flee from him . . . (Relig) (Anal-literal) (Animals/animals-MAMR) (INFERRED) (Inductive)
Contrapositive: verse? – NOT in Matthew, Mark, or Luke

1165. (if one) . . . came before me (then that one is) . . . (a) thief and (a) robber . . . (Relig) (MAMR) (INFERRED) (Deductive)
Contrapositive: verseF? – NOT in Matthew, Mark, or Luke

1166. (if one is a) . . . thief and (a) robber (then) the sheep did not hear (him) . . . (Relig) (MA/Animals-MAMR) (INFEREED) (Deductive)
Contrapositive: verse? – similar to # 1163, but not the same

1167. . . . if any man enter in . . . by me . . . (then) he shall be saved, and shall go in and out, and find pasture. (Relig) (MA-GA/MR) (Anal-s) (INFERRED-though "if" given) (Inductive)
Contrapositive: verse? – NOT in Matthew, Mark, or Luke

1168. (since) . . . I (Jesus) am come, (then) . . . they (the sheep) might have life, and that they might have it more abundantly. (Relig) (GAMR) (INFERRED) (Deductive)
Contrapositive: verse? – NOT in Matthew, Mark, or Luke

1169. (if one is) . . . the good shepherd, (then that one) . . . giveth his life for the sheep (Relig) (Anal-literal) (MAMR – GAMR) (INFERRED) (Deductive)
Contrapositive: verseF? – NOT in Matthew, Mark, or Luke

1170. (if one is) . . . an hireling, and not the shepherd, whose own the sheep are not, seeth the wolf coming, (then that one) leaveth the sheep, and fleeth: and the wolf catcheth them, and scattereth the sheep. (Relig) (Anal-literal) (MA/Animals-Satan/MR) (INFERRED) (Deductive)
Contrapositive: verse? – NOT in Matthew, Mark, or Luke

1171. (when that one who is an hireling and not the shepherd) leaveth the sheep, and fleeth (then) . . . the wolf catcheth them, and scattereth the sheep. (Relig) (Anal-literal) (MA/Animals-Satan/MR) (INFERRED) (Deductive)
Contrapositive: verseF? – NOT in Matthew, Mark, or Luke

1172. . . . because (one) is an hireling, and careth not for the sheep . . . the hireling fleeth (when the wolf cometh) . . . (Relig) (Anal-literal) (MAMR-Satan/Satan) (INFERRED) (Neither inductive nor deductive)
Contrapositive: verse? – NOT in Matthew, Mark, or Luke

1173. (if one is a sheep not of this fold then) . . . they (the sheep of the other fold) shall hear my voice . . . (Relig) (Anal-s) (MA/MR-GR) (INFERRED) (Inductive)
Contrapositive: verse? – true vaccuously

1174. (since) . . . I (Jesus) lay down my life that I might take it again . . . therefore doth my Father love me . . . (Relig) (GAGR) (INFERRED) (Neither inductive nor deductive)
Contrapositive: verse? – NOT in Matthew, Mark, or Luke

1175. (if one is a) . . . man (then that one cannot) . . . taketh my life from me . . . (Relig) (MAMR) (INFERRED) (Deductive)
Contrapositive: verse? – NOT in Matthew, Mark, or Luke

1176. . . . because ye are not of my sheep . . . (then) . . . ye believe not . . . (Relig) (Anal-s (MAMR) (INFERRED) (Deductive)
Contrapositive: verse? – NOT in Matthew, Mark, or Luke

1177. (if one is of) . . . my sheep (Jesus' sheep) (then that one) . . . will hear my voice, and I (Jesus) know (that one) . . . , and (that one) follow(s)

me: And I give unto (that one) . . . eternal life; and they shall never perish, neither shall any man pluck (that one) out of my hand. (Relig) (Anal-literal) (Animals/Animals – MAMR) (INFERRED) (Inductive)
Contrapositive: verseF? – NOT in Matthew, Mark, or Luke

1178. (if one is a) . . . man (any) (then that one) . . . is (not) able to pluck (the sheep of Jesus) . . . out of my Father's hand. (Relig) (Anal-s) (MA/MR-GR) (INFERRED) (Deductive)
Contrapositive: verse? – true vaccuously

1179. (if one is a) . . . man (any) (then that one) . . . is (not) able to pluck (the sheep of Jesus) . . . out of my hand. (Relig) (Anal-s) (MA/MR-GR) (INFERRED) (Deductive)
Contrapositive: verse? – true vaccuously

1180. If he called them gods, unto whom the word of God came, and (since) the scripture cannot be broken; (then) Say ye of him, whom the Father hath sanctified, and sent into the world, Thou blasphemest . . . (Relig) (GAMR) (INFERRED) (Deductive)
Contrapositive: verse? – NOT in Matthew, Mark, or Luke

1181. . . . (since) I (Jesus) said, I am the Son of God, (then) . . . Say ye of him, whom the Father hath sanctified, and sent into the world, Thou blasphemest . . . (Relig) (GAMR) (INFERRED) (Deductive)
Contrapositive: verse? – NOT in Matthew, Mark, or Luke

1182. If I (Jesus) do not the works of my Father, (then) believe me not . . . (Relig) (GAMR) (INFERRED – though "if" given) (Deductive)
Contrapositive: verse? – vaccuously true – NOT in Matthew, Mark, or Luke

1183. if I (Jesus) do (the works of my Father), though you believe not me, (then) believe the works . . . (Relig) (GA-MA/MR) (INFERRED-THOUGH "IF" GIVEN) (Deductive)
Contrapositive: verseF? – NOT in Matthew, Mark, or Luke; would be violateed in human terms by # 1272 ; see # 1045 and # 1396

1184. If any man walk in the day, (then) he stumbleth not . . . (Disc) (Anal-literal) (MAMR-MAMR) (INFERRED- though "if" given) (Deductive)
Contrapositive: verseF? – NOT in Matthew, Mark, or Luke

1185. (Since) . . . he (a man) seeth the light of this world, (then) he stumbleth not . . . (Disc) (Anal-literal) (MAMR-MAMR) (INFERRED- though "if" given) (Deductive)
Contrapositive: verseF? – NOT in Matthew, Mark, or Luke

1186. if a man walk in the night, (then) he stumbleth . . . (Disc) (Anal-literal) (MAMR-MAMR) (INFERRED – though "if" given) (Deductive)
Contrapositive: verseF? – NOT in Matthew, Mark, or Luke

1187. . . . (since) there is no light in him (a man), (then) he stumbleth . . . (Disc) (Anal-literal) (MAMR-MAMR) (INFERRED) (Deductive)
Contrapositive: verseF? – NOT in Matthew, Mark, or Luke

1188. (if one) . . . believeth in me, though he were dead, (then) yet shall (that one) . . . live. (Martha) (MA/GR-MR) (INFERRED) (Inductive)
Contrapositive: verse? – NOT in Matthew, Mark, or Luke

1189. (if one) . . . liveth and believeth in me, (then that one) shall never die. (Martha) (Anal-s) (INFERRED) (Inductive)
Contrapositive: verse? – NOT in Matthew, Mark, or Luke

1190. . . . because of the people which stand by, (then) I (Jesus) said it (I knew that thou hast heard me) . . . (Mult) (MAGR) (INFERRED) (Deductive)
Contrapositive: verseF? – NOT in Matthew, Mark, or Luke

1191. . . . because . . . I (Jesus) said it (I knew that thou hast heard me (then) they (the people) may believe that thou hast sent me (Mult) (GAMR) (INFERRED) (Deductive)
Contrapositive: verseF? – NOT in Matthew, Mark, or Luke

1192. . . . (since) the poor always ye have with you; but me ye have not always, (then) . . . Let her (Mary) alone . . . (Judas & Disciples and Mary's family) (MA-GA/MR) (INFERRED) (Neither inductive nor deductive)
Contrapositive: verseF? – NOT in Matthew, Mark, or Luke

1193. . . . (if) . . . a corn of wheat fall (not) into the ground and die, (then) it abideth alone . . . (Greeks, Phillip, Andrew) (Anal-literal) (GAGR-GAMR) (INFERRED) (Deductive)
Contrapositive: verseF? – NOT in Matthew, Mark, or Luke

1194. . . . (if) . . . a corn of wheat die, (then) it bringeth forth much fruit . . . (Greeks, Philip, Andrew) (Anal-literal) (GAGR-GAMR) (INFERRED) (Deductive)
Contrapositive: verseF? – NOT in Matthew, Mark, or Luke (similar to parable of sower, but no mention of "die" there)

1195. (if one) . . . loveth his life, (then that one) shall lose it (his life) . . . (Greeks, Philip, Andrew) (MAMR) (INFERRED) (Inductive)
Contrapositive: verseF? – NOT in Matthew, Mark, or Luke—similar to # 94, 165, and 471, but not the same as "loveth" is different than "save"

1196. (if one) . . . hateth his life in this world, (then that one) shall keep it (his life) unto life eternal . . . (Greeks, Philip, Andrew) (MAMR) (INFERRED) (Inductive)
Contrapositive: verseF? – NOT in Matthew, Mark, or Luke – similar to # 95, 166, and 472, but "hateth" is different than "lose"

1197. If any man serve me, (then) let him follow me . . . (Greeks, Philip, and Andrew) (MAMR) (INFERRED, but "if" is given) (Deductive)
Contrapositive: verse? – NOT in Matthew, Mark, or Luke

1198. (if) . . . I (Jesus) am (in a place), (then) there shall my servant be . . . (Greeks, Philip, and Andrew) (MAMR) (INFERRED) (Deductive)
Contrapositive: verseF? – as Jesus is everywhere – NOT in Matthew, Mark, or Luke

1199. If any man serve me, (then) him will my Father honour. (Greeks, Philip, and Andrew) (MAMR) (INFERRED, but "if" is given) (Inductive)
Contrapositive: verse? – NOT in Matthew, Mark, or Luke

1200. (since) this cause (occurred – that is, man's sins and God's Plan for salvation) (then) came I (Jesus) unto this hour. (Greeks, Philip, and Andrew) (MA-GA/GR) (INFERRED) (Deductive)
Contrapositive:verseF? – NOT in Matthew, Mark, or Luke

1201. (since it is) . . . Now (then at this time) . . . is the judgment of this world (occurring) . . . (Greeks, Philip, and Andrew) (GAGR) (INFERRED) (Deductive)
Contrapositive: verse? – NOT in Matthew, Mark, or Luke

1202. (since it is) . . . Now (then at this time) . . . shall the prince of this world be cast out. (Greeks, Philip, and Andrew) (GAGR) (INFERRED) (Deductive)
Contrapositive: verse? – NOT in Matthew, Mark, or Luke

1203. (if one) . . . ha(s) the light, (then) . . . Walk . . . (Greeks, Philip, and Andrew) (GAMR) (Anal-literal) (INFERRED) (Deductive)
Contrapositive: verseF? – NOT in Matthew, Mark, or Luke—see # 1121 for very close to converse since "walk" here has a meaning like "follow"; SEE #1205 FOR ESSENTIALLY THE CONTRAPOSITIVE

1204. (if one does not) . . . walk (in the light then) . . . . darkness (shall) come upon (that one) . . . (Greeks, Philip, and Andrew) (MAGR) (Anal-literal) (INFERRED) (Inductive)
Contrapositive: verse? – SEE #1204 FOR VERY ESSENTIALLY THE CONTRAPOSITIVE (ASSUMING THAT DARKNESS COME UPON ONE IS THE NEGATION OF HAVING THE LIGHT WHICH APPEARS TO BE THE ASSUMPTION OF CHRIST SUCH AS IN # 1121 WHICH HAS THE SAME CONTEXT OF "WALK")

1205. (since) he that walketh in darkness knoweth not whither he goeth (then that one will have) darkness come upon (that one) . . . . (Greeks, Philip, and Andrew) (MAGR) (Anal-literal) (INFERRED) (Inductive)
Contrapositive: verseF? – NOT in Matthew, Mark, or Luke

1206. (if one) . . . walketh in darkness (then that one) knoweth not whither he goeth . . . (Greeks, Philip, and Andrew) (MAMR) (Anal-literal) (INFERRED) (Deductive)
Contrapositive: verseF? – NOT in Matthew, Mark, or Luke

1207. While ye have the light, (then) believe in the light . . . (Greeks, Philip, and Andrew) (GAMR) (INFERRED) (Deductive)
Contrapositive: verseF? – NOT in Matthew, Mark, or Luke

1208. . . . (if ye) . . . believe in the light, (then) . . . ye may be the children of light. (Greeks, Philip, and Andrew) (MA/MR-GR) (INFERRED) (Deductive)
Contrapositive: verse? – NOT in Matthew, Mark, or Luke

1209. (if one) . . . believeth on me, (then that one) . . . believeth not on me . . . (General – not specific) (MA-GA/MR-GR) (INFERRED) (Deductive)
Contrapositive: verseF? – NOT in Matthew, Mark, or Luke

1210. (if one) . . . believeth on me, (then that one) . . . believeth . . . on him that hath sent me . . . (General – not specific) (MA-GA/MR-GR) (INFERRED) (Deductive)
Contrapositive: verse? – NOT in Matthew, Mark, or Luke

1211. (if one) . . . seeth me, (then that one) seeth him that hath sent me. (General – not specific) (MA-GA/MR-GR) (INFERRED) (Deductive)
Contrapositive: verseF? – NOT in Matthew, Mark, or Luke

1212. (since) . . . I am come a light into the world, (then) . . . whosoever believeth on me should not abide in darkness. (General – not specific) (MA-GA/MR-GR) (INFERRED) (Deductive)
Contrapositive: verseF? – NOT in Matthew, Mark, or Luke

1213. (if one) . . . believeth on me (then that one) should not abide in darkness. (General – not specific) (MA-GA/MR-GR) (INFERRED) (Deductive)
Contrapositive: verse? – NOT in Matthew, Mark, or Luke—see # 1121 for similar, but "followeth" is not exactly the same as "believe"

1214. . . . if any man hear my words, and believe not, (then) I judge him not . . . (General-not specific) (MAGR) (INFERRED) (Deductive)
Contrapositive: verse? – Not in Matthew, Mark, or Luke—see # 1056 for hear and believe with opposite conclusion

1215. . . . since I (Jesus) came not to judge the world, but to save the world, (then) I judge him (one who hears, but does not believe) not . . . (General audience) (MAGR) (INFERRED) (Deductive)
Contrapositive: verse? – NOT in Matthew, Mark, or Luke—see #1028 for essentially the same condition, but a different conclusion—see also # 1215 for same conclusion with "since" to explain

1216. (if one) . . . rejecteth me, and receiveth not my words, (then that one) hath one that judgeth him . . . . (General-not specific) (MA/MR-GR) (INFERRED) (Deductive)
Contrapositive: verse? – NOT in Matthew, Mark, or Luke

1217. (when it is) . . . the last day (then) . . . the word that I have spoken . . . shall judge (that one who rejecteth Christ) (General audience) (GA/GR-MR) (INFERRED) (Deductive)
Contrapositive: verse? – NOT in Matthew, Mark, or Luke

1218. (since) . . . I have not spoken of myself; but the Father which sent me, (then) . . . the word that I have spoken, the same shall judge him in the last day. (General audience) (GA/GR-MR)
Contrapositive: verse? – NOT in Matthew, Mark, or Luke

1219. (if) I (Jesus) speak therefore as the Father said unto me, so I speak. (General audience) (GAGR) (INFERRED) (Deductive)
Contrapositive: verse?—see second half of # 1129 for similar, but not same

1220. if I wash thee (Peter) not, (then) thou hast no part with me. (Peter) (GAMR) (INFEERED- though "if" given) (Neither deductive nor inductive)
Contrapositive: verse? – NOT in Matthew, Mark, or Luke

1221. (if one) . . . is washed (then that one) needeth not save to wash his feet, but is clean every whit . . . (Peter) (Anal-literal) (MAMR-GAMR) (INFERRED) (Deductive)
Contrapositive: verse? – NOT in Matthew, Mark, or Luke

1222. (since) . . . I (Jesus) am (Master and Lord) (then) . . . ye say well . . . (Disc) (GAMR) (INFERRED) (Deductive)
Contrapositive: verseF? – Jesus is Master and Lord regardless – NOT in Matthew, Mark, or Luke

1223. If I (Jesus) then, your Lord and Master, have washed your feet (then) ye also ought to wash one another's feet. (Disc) (GAMR) (INFERRED – though "if" given) (Deductive)
Contrapositive: verseF? – NOT in Matthew, Mark, or Luke

1224. (since) . . . I (Jesus) have given you an example, (then) . . . ye also ought to wash one another's feet. (Disc) (GAMR) (INFERRED) (Deductive)
Contrapositive: verseF? – NOT in Matthew, Mark, or Luke

1225. (if one is) . . . a servant (then that one) is not greater than his lord . . . . (Disc) (Anal-literal) (MAMR-MAGR) (INFERRED) (Deductive)
Contrapositive: verse? – NOT in Matthew, Mark, or Luke

1226. (if one is) . . . sent (then that one) is not greater than he that sent him. (Disc) (Anal-literal) (MAMR-MAGR) (INFERRED) (Deductive)
Contrapositive: verse? – NOT in Matthew, Mark, or Luke

1227. If ye know these things (about servanthood) . . . and do them (then) . . . happy are ye . . . (Disc) (MA-GA/MR-GR) (INFERRED) (Deductive)
Contrapositive: verse?—see # 746 for similar, but NOT in Matthew, Mark, or Luke

1228. . . . before it come . . . Now I tell you . . . (Disc) (GAGR) (INFERRED) (Neither inductive nor deductive)
Contrapositive: verse? – NOT in Matthew, Mark, or Luke

1229. Now I tell you before it come (then) . . . when it is come to pass, ye may believe that I (Jesus) am he. (Disc) (GA/GR-MR) (INFERRED) (Neither inductive nor deductive)
Contrapositive: verseF? – NOT in Matthew, Mark, or Luke

1230. when it is come to pass, (then) ye may believe that I am he. (Disc) (GA/GR-MR) (INFERRED) (Neither inductive nor deductive)
Contrapositive: verseF? – NOT in Matthew, Mark, or Luke

1231. (if one) . . . receiveth whomsoever I send (then that one) receiveth me . . . (Disc) (MA-GA/MR) (INFERRED) (Deductive)
Contrapositive: verse?—see # 96 for special case of condition with disciples and same conclusion – NOT in Matthew, Mark, or Luke

1232. see # 97 for same

1233. when I have dipped it (the sop) (then) . . . the one to whom I shall give a sop . . . is . . . he (whom shall betray me). (John the disciple) (GA/GR-MR) (INFERRED) (Neither inductive nor deductive)
Contrapositive: verseF? – NOT in Matthew, Mark, or Luke

1234. (since) . . . now . . . (is the time that Jesus is betrayed) . . . the Son of man is glorified, and God is glorified in him. (Disc) (GAGR) (INFERRED) (Deductive)
Contrapositive: verseF? – NOT in Matthew, Mark, or Luke

1235. If God be glorified in him, (then) God shall also glorify him in himself, and shall straightway glorify him. (Disc) (GAGR) (INFERRED – though "if" given) (Deductive)
Contrapositive: verse? – NOT in Matthew, Mark, or Luke

1236. . . . yet a little while, (then) . . . Little children, . . . I am with you . . . (Disc) (GA/GR-MR) (INFERRED) (Neither inductive nor deductive)
Contrapositive: verse? – NOT in Matthew, Mark, or Luke

1237. (since) . . . Whither I go, ye cannot come (then) so now I say unto you. A new commandment I give . . . (Disc) (GA-MR/GR) (INFERRED) (Deductive)
Contrapositive: verse? – NOT in Matthew, Mark, or Luke

1238. (since) as I have loved you, (then) ye also love one another. (Disc) (GAMR) (INFERRED) (Deductive)
Contrapositive: verseF? – NOT in Matthew, Mark, or Luke

1239. . . . if ye have love one for another . . . (then) all men shall know that ye are my disciples . . . (Disc) (MAMR) (INFERRED- though "if" given) (Inductive)
Contrapositive: verseF? – assumes all men have contact and knowledge of the disciple – NOT in Matthew, Mark, or Luke

1240. . . . afterwards (after Jesus goes away by death) . . . (then) thou (the disciples) shalt follow me (Jesus) . . . (Disc) (GA/MR-GR) (INFERRED) (Deductive)
Contrapositive: verse? – NOT in Matthew, Mark, or Luke

1241. (un)till thou (Peter) has denied me thrice (then) the cock shall not crow (Peter) (MAGR) (INFERRED) (Neither inductive nor deductive)
Contrapositive: verse? - very similar to #383 and #993, but this has condition on Peter's denial, not the cock crowing

1242. (since) ye believe in God, (then) believe also in me (Jesus). (Disc) (MA-GA/MR-GR) (INFERRED) (Deductive)
Contrapositive: verseF? – man can believe in God but not that Jesus is God's Son as many do – NOT in Matthew, Mark, or Luke

1243. . . . if it (the mansions in my Father's house) were not so, (then) I (Jesus) would have told you. (Disc) (GAGR) (INFERRED) (Deductive)
Contrapositive: verse? – NOT in Matthew, Mark, or Luke

1244. . . . if I (Jesus) go and prepare a place for you, (then) I (Jesus) will come again, and receive you unto myself . . . (Disc) (GA/GR-MR) (INFERRED – though "if" given) (Deductive)
Contrapositive: verse? – NOT in Matthew, Mark, or Luke

1245. (if) I (Jesus) will come again, and receive you unto myself (then) where I am, there ye may be also. (Disc) (GA/GR-MR) (INFERRED – though "if" given) (Deductive)
Contrapositive: verse? – NOT in Matthew, Mark, or Luke

1246. where I (Jesus) am, there ye may be also. (Disc) (GA/GR-MR) (INFERRED – though "if" given) (Deductive)
Contrapositive: verseF? – Jesus is in heaven now according to the Bible, but we are not there if we are alive on the earth – NOT in Matthew, Mark, or Luke

1247. (if one is a) . . . man (then that one) cometh (not) unto the Father, but by me (Jesus). (Disc) (GA/MR-GR) (INFERRED) (Deductive)
Contrapositive: verseF? (condition is always true, but conclusion must be false– NOT in Matthew, Mark, or Luke

1248. If ye had known me, (then) ye should have known my Father also . . . (Disc) (MAMR) (INFERRED- though "if" given) (Deductive)
Contrapositive: verse?—see # 1125 for same; see # 1055 for somewhat like inverse, though "known" and "honoureth" are different

1249. . . . from henceforth (then) ye know him (God the Father), and have seen him. (Disc) (GAMR) (INFERRED) (Deductive)
Contrapositive: verse? – NOT in Matthew, Mark, or Luke – NOT the same as # 1025—see # 1025 for counterexample in which disciples "from henceforth" have seen the Father (same Greek word for "seen" since this statement is not qualified in time

1250. (if one) . . . hath seen me (then that one) hath seen the Father . . . (Disc) (MA-GA/MR-GR) (INFERRED) (Deductive)
Contrapositive: verseF? – – NOT in Matthew, Mark, or Luke
see # for counterexample in which disciples "from henceforth" have seen the Father (same Greek word for "seen" since this statement is not qualified in time

1251. (if one) . . . believeth on me (then) the works I do shall he do also; and greater works than these shall he do . . . (Disc) (MA-GA/MR-GR) (INFERRED) (Inductive)
Contrapositive: verse? – NOT in Matthew, Mark, or Luke

1252. because (since) I (Jesus) go to my Father (if one) . . . believeth on me (then) the works I do shall he do also; and greater works than these shall he do . . . (Disc) (GA/MR-GR) (INFERRED) (Inductive)
Contrapositive: verse? - – NOT in Matthew, Mark, or Luke

1253. (if) ye shall ask (anything) in my name, (then) . . . I (Jesus) . . . will . . . do (that thing) . . . (Disc) (MAGR) (INFERRED) (Deductive)
Contrapositive: verseF? – Christ does things without us necessarily asking, even for believers – NOT in Matthew, Mark, or Luke, but see # 233 for very similar with part of conditions different

1254. (if) ye shall ask (anything) in my name, (then) . . . I . . . will . . . do (that thing) . . . (then) the Father may be glorified in the Son. (Disc) (MA-GA/GR) (INFERRED) (Deductive)
Contrapositive: verseF? – the Father can be glorified independent of man's actions – NOT in Matthew, Mark, or Luke

1255. Same as # 1254

1256. If ye love me, (then) keep my commandments. (Disc) (MAMR) (INFERRED – though "if" given) (Deductive)
Contrapositive: verse?—see John 14: 21 (#1265) for essentially converse; and John 14: 24 for inverse – NOT in Matthew, Mark, or Luke

1257. (when) . . . I will pray the Father, (then) . . . he shall give you another Comforter . . . (Disc) (GA/GR-MR) (INFERRED) (Deductive)
Contrapositive: verse? – NOT in Matthew, Mark, or Luke

1258. (when) . . . I will pray the Father, and he shall give you another Comforter, (then) he may abide with you for ever. (Disc) (GA/GR-MR) (INFERRED) (Deductive)
Contrapositive: verseF? – would seem to false as the Spirit may not abide with us, but Jesus could have prayed and the Father gave us another Comforter – NOT in Matthew, Mark, or Luke

1259. . . . because (since) it (the world) seeth him (the Spirit of truth) not, neither knoweth him, (then) . . . Even the Spirit of truth whom the world cannot receive . . . (Disc) (MAMR) (INFERRED) (Deductive)
Contrapositive: verse? – NOT in Matthew, Mark, or Luke

1260. . . . (since) he (the Spirit of truth) dwelleth with you, and shall be with you . . . (then) ye (the disciples) know him . . . (Disc) (GAMR) (INFERRED) (Deductive)
Contrapositive: verse? – NOT in Matthew, Mark, or Luke

1261. Yet a little while, (then) . . . the world seeth me no more, but ye see me . . . (Disc) (GAMR) (INFERRED) (Deductive)
Contrapositive: verseF? – NOT in Matthew, Mark, or Luke

1262. . . . because I (Jesus) live, (then) ye shall live also. (Disc) (GAMR) (INFERRED) (Deductive)
Contrapositive: verseF? – false if one assumes Jesus lives – NOT in Matthew, Mark, or Luke

1263. (When) . . . At that day (the world seeth me no more), (then) ye shall know that I am in my Father, and ye in me, and I in you. (Disc) (GAMR) (INFERRED) (Deductive)
Contrapositive: verseF? – Peter's knowledge of who Jesus is before that day – NOT in Matthew, Mark, or Luke

1264. (if one) hath my commandments, and keepeth them . . . (then that one) loveth me . . . (Disc) (MAMR) (INFERRED) (Deductive)
Contrapositive: verseF? – NOT in Matthew, Mark, or Luke – This is the contrapositive of

1265. (if one) . . . loveth me (then that one) shall be loved of my Father, and I will love him, and will manifest myself to him. (Disc) (MAGR) (INFERRED) (Inductive)
Contrapositive: verse? – NOT in Matthew, Mark, or Luke

1266. If a man love me, (Jesus) (then) he (that man) will keep my words: and my Father will love him, and we will come unto him, and make our abode with him. (Disc) (MA/MR-GR) (INFERRED) (Inductive)
Contrapositive: verse? – NOT in Matthew, Mark, or Luke – similar to # 1256 (extra conclusion); also very close to converse of # 1264 (except for "hath my commandments"); and very close to inverse of # 1267

1267. (if one) loveth me not . . . (then that one) . . . keepeth not my sayings . . . (Disc) (MAMR) (INFERRED) (Deductive)
Contrapositive: verseF? – NOT in Matthew, Mark, or Luke – This is Inverse of #1266 and most importantly, the contrapositive of # 1264 (though "sayings" is used in #1267 as opposed to "commands" in # 1264)
STINV

1268. . . . being yet present with you . . . (then) . . . These things have I spoken unto you . . . (Disc) (GAGR) (INFERRED) (Neither inductive nor deductive)
Contrapositive: verseF? – NOT in Matthew, Mark, or Luke

1269. If ye loved me (then) ye would rejoice . . . (Disc) (MAMR) (INFERRED) (Deductive)
Contrapositive: verseF? – NOT in Matthew, Mark, or Luke

1270. . . . because (since) I said, I (Jesus) go to my Father . . . (then) If ye loved me (then) ye would rejoice . . . (Disc) (GA-MA/MR) (INFERRED) (Deductive)
Contrapositive: verseF? – NOT in Matthew, Mark, or Luke

1271. . . . (since) my Father is greater than I (then) . . . because (since) I said, I (Jesus) go to my Father . . . (then) If ye loved me (then) ye would rejoice . . . (Disc) (GA-MA/MR) (INFERRED) (Deductive)
Contrapositive: verseF? – NOT in Matthew, Mark, or Luke—see #1227 for more general principle – SEE # 1412 and # 1413 for why contrapositive would be false since the conclusion of the contrapositive could not be true given # 1412 and #1413, according to the strict rules of formal, human reasoning

1272. (since) . . . I have told you before it (when Jesus goes to the Father) come to pass . . . (then) when it is come to pass, ye might believe. (Disc) (GAMR) (INFERRED) (Deductive)
Contrapositive: verse?—see #1230 - essentially similar, but different “it”

1273. see #1231 for essentially similar, but different “it”

1274. . . . before it come to pass . . . I have told you . . . (Disc) (GAGR) (INFERRED) (Deductive)
Contrapositive: verse?—see #1229 - essentially similar, but different “it”

1275. Hereafter (then) I (Jesus) will not talk much with you . . . (Disc) (GAGR) (INFERRED) (Deductive)
Contrapositive: verse? – NOT in Matthew, Mark, or Luke

1276. . . . (since) the prince of this world cometh, and hath nothing in me (then) . . . Hereafter I (Jesus) will not talk much with you . . . (Disc) (Satan-GA/GR) (INFERRED) (Deductive)
Contrapositive: verse? – NOT in Matthew, Mark, or Luke

1277. . . . as (since) the Father gave me commandment, (then) .. even so I (Jesus) do. (Disc) (GAGR) (INFERRED) (Deductive)
Contrapositive: verse? – NOT in Matthew, Mark, or Luke

1278. . . . as (since) the Father gave me commandment, even so I (Jesus) do . . . (then) the world may know that I love the Father. (Disc) (GAMR) (INFERRED) (Deductive)
Contrapositive: verse? – NOT in Matthew, Mark, or Luke

1279. (if a) branch in me beareth not fruit (then) he (the Father) taketh away (the branch) . . . (Disc) (MAGR) (Anal-s) (INFERRED) (Deductive)
Contrapositive: verseF? – NOT in Matthew, Mark, or Luke

1280. (if a) branch in me beareth fruit (then) he (the Father) purgeth it (the branch) . . . (Disc) (Anal-s) (MAGR) (INFERRED) (Deductive)
Contrapositive: verseF? – NOT in Matthew, Mark, or Luke

1281. (if a) branch in me beareth fruit (then) he (the Father) purgeth it (the branch) . . . (then) it (the branch) may bring forth more fruit. (Disc) (Anal-s) (MAGR) (INFERRED) (Deductive)
Contrapositive: verseF? – NOT in Matthew, Mark, or Luke

1282. Now (then) ye are clean through the word which I have spoken unto you. (Disc) (Anal-s) (GA/MR-GR) (INFERRED) (Deductive)
Contrapositive: verse? – NOT in Matthew, Mark, or Luke

1283. As (since) the branch cannot bear fruit of itself, except it abide in the vine (then) no more ye (cannot bear fruit of yourself, except ye abide in me (Jesus), except ye abide in me. (Disc) (Anal-literal) (Branch/Branch and MAMR) (INFERRED) (Deductive)
Contrapositive: verse? – NOT in Matthew, Mark, or Luke

{
1284. (if) . . . ye (do not) abide in me (Jesus), (then) . . . ye (cannot bear fruit). (Disc) (Anal-literal) (MA/MR-GR) (INFERRED) (Deductive)
Contrapositive: verse? – NOT in Matthew, Mark, or Luke

1285. (if one) . . . abideth in me (Jesus), and I in him (then that one) . . . bringeth forth much fruit . . . (Disc) (MA/MR-GR) (Anal-s) (INFERRED) (Deductive)
Contrapositive: verse? – NOT in Matthew, Mark, or Luke—very close to STINV with # 1285}

1286. . . . (since) without me ye can do nothing (then that one who abideth in me) . . . bringeth forth much fruit . . . (Disc) (MA/MR-GR) (Anal-s) (INFERRED) (Deductive)
Contrapositive: verse? –NOT in Matthew, Mark, or Luke

1287. (if one is) without me (then that one) . . . can do nothing (to bring forth fruit) . . . (Disc) (MA/MR-GR) (INFERRED) (Deductive)
Contrapositive: verseF? – NOT in Matthew, Mark, or Luke

1288. If a man abide not in me, (then) he is cast forth as a branch, and is withered; and men gather them, and cast them into the fire, and they are burned. (Disc) (Anal-s) (MA/MR-GR) (INFERRED – though "if" given) (Deductive)
Contrapositive: verse? – NOT in Matthew, Mark, or Luke

1289. If ye abide in me, and my words abide in you, (then) ye shall ask what ye will, and it shall be done unto you. (Disc) (MA/MR-GR) (INFERRED) (Inductive)
Contrapositive: verse? – NOT in Matthew, Mark, or Luke

1290. (if) ye shall ask what ye will, and it shall be done unto you . . . (then) Herein is my Father glorified . . . (Disc) (MA-GR/GR) (INFERRED) (Deductive)
Contrapositive: verseF? – NOT in Matthew, Mark, or Luke—see # 1255 for very similar, but there the "Father is glorified in the Son"

1291. (if) ye shall ask what ye will, and it shall be done unto you . . . (then) ye bear much fruit. (Disc) (MA-GR/MR) (INFERRED) (Deductive)
Contrapositive: verse? – NOT in Matthew, Mark, or Luke

1292. (Since) . . . the Father loved me, (then) so have I (Jesus) loved you . . . (Disc (GAGR) (INFERRED) (Deductive)
Contrapositive: verse? – NOT in Matthew, Mark, or Luke

1293. Essentially similar to # 1265, but not the same

1294. . . . if ye do whatsoever I command you (then) . . . ye are my friends . . . (Disc) (MAMR) (INFERRED) (Deductive)
Contrapositive: verse? – NOT in Matthew, Mark, or Luke

1295. Henceforth (then) I call you not servants . . . (Disc) (GAGR) (INFERRED) (Deductive)
Contrapositive: verse? – NOT in Matthew, Mark, or Luke

1296. . . . (since) the servant knoweth not what his lord doeth . . . Henceforth I call you not servants . . . (Disc) (GAGR) (INFERRED) (Deductive)
Contrapositive: verse? – NOT in Matthew, Mark, or Luke

1297. . . . (since) all things that I have heard of my Father I have made known unto you (then) I have called you friends . . . (Disc) (GAGR) (INFERRED) (Deductive)
Contrapositive: verse? – NOT in Matthew, Mark, or Luke

1298. (Since) . . . Ye have not chosen me, but I (Jesus) have chosen you, and ordained you, (then) ye should go and bring forth fruit, and that your fruit should remain . . . (Disc) (GAMR) (INFERRED) (Deductive)
Contrapositive: verseF? – NOT in Matthew, Mark, or Luke

1299. (Since) . . . Ye have not chosen me, but I (Jesus) have chosen you, and ordained you, (then) ye should go and bring forth fruit, and that your fruit should remain (then) whatsoever ye shall ask of the Father in my name, he may give it you. (Disc) (GA-MA/MR-GR) (INFERRED) (Inductive)
Contrapositive: verse? – NOT in Matthew, Mark, or Luke—Conclusion is found essentially the same in # 1254
1300a) (if) ye shall ask of the Father (anything) in my name, (then) he (the Father) will give it unto you. (Disc) (MAGR) (INFERRED) (Deductive)

1300. If the world hate you, (then) ye know that it hated me before it (the world) hated you. (Disc) (MAMR) (INFERRED) (Deductive)
Contrapositive: verseF? – we need not know things for them to be true – NOT in Matthew, Mark, or Luke

{
1301. If ye were of the world, (then) the world would love his own . . . (Disc) (MAMR) (INFERRED) (Deductive)
Contrapositive: verseF? – NOT in Matthew, Mark, or Luke

1302. . . . because ye are not of the world, but I have chosen you out of the world, therefore the world hateth you. (Disc) (MA-GA/MR) (NOT INFERRED) (Deductive)
Contrapositive: verse? – close to STINV (love and hateth are not exact negations) – NOT in Matthew, Mark, or Luke
}

1303. If they have persecuted me, (then) they will also persecute you . . . (Disc) (MAMR) (INFERRED- though ‘if’ given) (Deductive)
Contrapositive: verseF? – NOT in Matthew, Mark, or Luke

1304. . . . if they have kept my saying, (then) they will keep yours also. (Disc) (MAMR) (INFERRED – though ‘if’ given) (Deductive)
Contrapositive: verseF? – NOT in Matthew, Mark, or Luke

1305. . . . (because of) my (Jesus’) name sake, (and) because they (the world) know not him that sent me . . . (then) all these things (persecutions) will they do unto you . . . (Disc) (GAMR) (INFERRED) (Inductive)
Contrapositive: verseF? – NOT in Matthew, Mark, or Luke

{

1306. If I (Jesus) had not come and spoken unto them (the world), (then) they had not had sin . . . (Disc) (GAMR) (INFERRED- though ‘if’ given) (Deductive)
Contrapositive: verseF? – NOT in Matthew, Mark, or Luke

1307. . . . now (that I (Jesus) have come and spoken unto them) (then) they have no cloak for their sin. (Disc) (GAMR) (INFERRED) (Deductive)
Contrapositive: verseF? – NOT in Matthew, Mark, or Luke – close STINV

}

1308. (If one) . . . hateth me (then that one) hateth the Father also. (Disc) (MAMR) (INFERRED) (Deductive)
Contrapositive: NOT in Matthew, Mark, or Luke—see # 1055 for similar, but not the same since “hateth” and “not honoureth” are different

1309. If I (Jesus) had not done among them what no one else did, (then) they (the world) would not be guilty of sin. (Disc) (GAMR) (INFERRED- though ‘if’ given) (Deductive)
Contrapositive: verseF? – NOT in Matthew, Mark, or Luke

1310. (Since) . . . now (that they have seen these miracles), (then) have they (the world) both seen and have hated both me and my Father. (Disc) (MA-GA/MR) (INFERRED) (Deductive)
Contrapositive: verseF? – NOT in Matthew, Mark, or Luke

1311. (since) . . . this (the world seeing the miracles of Jesus, but hating him and the Father) cometh to pass (then) . . . the word might be fulfilled that is written in their law, They hated me without a cause. (Disc) (GA-MR/GR-MR) (INFERRED) (Deductive)
Contrapositive: verse? – NOT in Matthew, Mark, or Luke

1312. . . . when the Comforter is come, whom I will send unto you from the Father, even the Spirit of truth, which proceedeth from the Father, (then) he (the Comforter) shall testify of me. (Disc) (GAGR) (INFERRED) (Neither inductive nor deductive)
Contrapositive: verse? – NOT in Matthew, Mark, or Luke

1313. . . . because ye have been with me from the beginning, (then) . . . ye also shall bear witness . . . (Disc) (MAMR) (INFERRED) (Neither inductive nor deductive)
Contrapositive: verse? – NOT in Matthew, Mark, or Luke

1314. (since) these things have I (Jesus) spoken unto you, (then) . . . ye should not be offended. (Disc) (GAMR) (INFERRED) (Deductive)
Contrapositive: verseF? – NOT in Matthew, Mark, or Luke

1315. (if one) killeth you (the disciples) (then that one) will think that he doeth God service. (Disc) (MAMR) INFERRED) (Inductive)
Contrapositive: verseF? – NOT in Matthew, Mark, or Luke

1316. . . . because they have not known the Father, nor me (Jesus) (then) . . . these things (killing the disciples) will they do unto you . . . (Disc) (MAMR) (INFERRED) (Inductive)
Contrapositive: verse? – NOT in Matthew, Mark, or Luke—similar to # 1306 (but additional condition)

1317. . . . when the time shall come (for the world's attempts to kill them) (then) ye may remember that I (Jesus) told you of them. (Disc) (GAMR) (INFERRED) (Neither inductive nor deductive)
Contrapositive: verseF? - NOT in Matthew, Mark, or Luke

1318. . . . because I (Jesus) was with you . . . these things (about the world's eventual attempts to kill the disciples) I said not unto you at the beginning (of Jesus' ministry)
(Disc) (GAGR) (NOT INFERRED) (Neither inductive nor deductive)
Contrapositive: verse? – NOT in Matthew, Mark, or Luke

1319. . . . because I (Jesus) have said these things unto you, (then) sorrow hath filled your heart. (Disc) (GAMR) (INFERRED) (Deductive)
Contrapositive: verseF? – NOT in Matthew, Mark, or Luke

1320. . . . (since) if I (Jesus) go not away, the Comforter will not come unto you . . . (then) It is expedient for you that I go away . . . (Disc) (GAMR) (INFERRED) (Deductive)
Contrapositive: verseF? – NOT in Matthew, Mark, or Luke

{
1321. . . . if I (Jesus) go not away, (then) the Comforter will not come unto you . . . (Disc) (GAGR) (INFERRED- though 'if' given) (Deductive)
Contrapositive: verse? – NOT in Matthew, Mark, or Luke

1322. . . . if I (Jesus) depart, (then) I (Jesus) will send him (the Comforter) unto you. (Disc) (GAGR) (INFERRED – though 'if' given) (Neither inductive nor deductive)

Contrapositive: verse? – NOT in Matthew, Mark, or Luke
} close to STINV

1323. . . . when he (the Comforter) is come, (then) he will reprove the world of sin, and of righteousness, and of judgment. (Disc) (GAGR) (INFERRED – though 'when' given) (Neither inductive nor deductive)
Contrapositive: verse? – NOT in Matthew, Mark, or Luke

1324. . . . because (since) they (the world) believe not on me . . . (then the Comforter will reprove the world) of sin . . . (Disc) (MAGR) (INFERRED) (Deductive)
Contrapositive: verse? – NOT in Matthew, Mark, or Luke

1325. . . . because (since) I (Jesus) go to my father, and ye see me no more . . . (then the Comforter will reprove the world) of righteousness . . . (Disc) (MAGR) (INFERRED) (Deductive)
Contrapositive: verse? – NOT in Matthew, Mark, or Luke

1326. . . . because (since) the prince of this world is judged . . . (then the Comforter will reprove the world) of judgment . . . (Disc) (MAGR) (INFERRED) (Deductive)
Contrapositive: verse? – NOT in Matthew, Mark, or Luke

1327. . . . now (prior to Jesus' Death) . . . ye (the discples) cannot bear them (many things which Jesus had yet to say) . . . (Disc) (GAMR) (INFERRED) (Neither inductive nor deductive)
Contrapositive: verse? – NOT in Matthew, Mark, or Luke

1328. . . . when he, the Spirit of truth, is come, (then) he will guide you into all truth . . . (DISC) (GAGR) (INFERRED) (Deductive)
Contrapositive: verse? – NOT in Matthew, Mark, or Luke

1329. . . . (since) he (the Spirit of truth) shall not speak of himself; but whatsoever he shall hear, that shall he speak . . . (then) when he, the Spirit of truth, is come, (then) he will guide you into all truth . . . (Disc) (GAGR) (INFERRED) (Deductive)
Contrapositive: verse? – NOT in Matthew, Mark, or Luke

1330. (if) he (the Spirit of truth) shall hear (anything), (then) shall he speak (that thing) . . . (Disc) (GAGR) (INFERRED) (Neither inductive nor deductive)
Contrapositive: verse? – NOT in Matthew, Mark, or Luke

1331. . . . (since) he (the Spirit of truth) shall receive of mine, and shall shew it unto you . . . (then) He shall glorify me . . . (Disc) (GAGR) (INFERRED) (Neither inductive nor deductive)
Contrapositive: verse? – NOT in Matthew, Mark, or Luke

1332. (If) . . . the Father hath (anything) (then that thing is) . . . mine (Jesus) . . . (Disc) (GAGR) (INFERRED) (Neither inductive nor deductive)
Contrpaositive: verse? – NOT in Matthew, Mark, or Luke—see # 1366 for same

1333. (since) All things that the Father hath are mine: therefore said I (Jesus), that he (the Spirit of truth) shall take of mine, and shall shew it unto you. (Disc) (GAGR) (INFERRED- though 'therefore' given) (Neither inductive nor deductive)
Contrapositive: verse? – NOT in Matthew, Mark, or Luke

1334. (For) A little while, (then) . . . ye shall not see me . . . (Disc) (GAMR) (INFERRED) (Neither inductive nor deductive)
Contrapositive: verse? – NOT in Matthew, Mark, or Luke

1335. (For) again (after the other 'little while' in # 1335), a little while, (then) . . . ye shall see me . . . (Disc) (GAMR) (INFERRED) (Neither inductive nor deductive)
Contrapositive: verse? – NOT in Matthew, Mark, or Luke – The word "again" is key to having this be not a violation to # 1335

1336. . . . because (since) I (Jesus) go to my Father . . . (then) again, a little while, and ye shall see me . . . (Disc) (GAMR) (INFERRED) (Neither inductive nor deductive)
Contrapositive: verse? – The word "shall" in # 1337 is important to this not in conflict with # 1326

1337. . . . when she (a woman having a child) is in travail (then she) . . . hath sorrow . . . (Disc) (GAMR) (INFERRED) (Deductive)
Contrapositive: verseF? – NOT in Matthew, Mark, or Luke

1338. . . . because her (a woman having a child) hour is come . . . (then) A woman when she is in travail hath sorrow . . . (Disc) (GA/MR-GR) (INFERRED) (Deductive)
Contrapositive: verse? – NOT in Matthew, Mark, or Luke

1339. . . . as soon as she is delivered of the child, (then) she remebereth no more the anguish, for joy that a man is born into the world. (Disc) (GA/MR-GR) (INFERRED) (Deductive)
Contrapositive: verseF? – NOT in Matthew, Mark, or Luke

1340. . . . (since there is) joy that a man is born into the world . . . as soon as she is delivered of the child, (then) she remebereth no more the anguish . . . (Disc) (GAMR) (INFERRED) (Deductive)
Contrapositive: verse? – NOT in Matthew, Mark, or Luke

1341. . . . now (since # 1335 and # 1337) . . . ye . . . therefore have sorrow . . . (Disc) (GAMR) (INFERRED) (Deductive)
Contrapositive: verseF? – NOT in Matthew, Mark, Luke

1342. (when) . . . I (Jesus) see you again, (then) . . . your heart shall rejoice, and your joy no man taketh from you. (Disc) (GAMR) (INFERRED) (Deductive)
Contrapositive: verse? – NOT in Matthew, Mark, Luke

1343. (if one is a) . . . man (and when) I (Jesus) see you again, (then) your joy (that man cannot) taketh from you. (Disc) (MA-GA/GR) (INFERRED) (Deductive)
Contrapositive: verse? – NOT in Matthew, Mark, Luke

1344. . . . in that day (when I see you again before I go to my Father), (then) ye shall ask me nothing . . . (Disc) (GAMR) (INFERRED) (Neither inductive nor deductive)
Contrapositive: verse? – NOT in Matthew, Mark, Luke

1345. same as # 1300a
1346. (up to the time just before Jesus' Death) Hitherto (then) have ye asked nothing in my name . . . (Disc) (GAMR) (INFERRED) (Neither inductive nor deductive)
Contrapositive: verse? – NOT in Matthew, Mark, Luke

1347. . . . (if ye) ask, (then) ye shall receive . . . (Disc) (MAGR) (INFERRED) (Inductive)
Contrapositive: verse? – NOT in Matthew, Mark, Luke (though # 60 is very similar)

1348. . . . (if ye) ask, (then) ye shall receive . . . (then) your joy may be full. (Disc) (MAGR) (INFERRED) (Inductive)
Contrapositive: verse? – NOT in Matthew, Mark, Luke

1349. . . . the time cometh when I (Jesus) shall no more speak unto you in proverbs, (then) . . . I (Jesus) shall shew you plainly of the Father. (Disc) (GAGR) (INFERRED) (Neither inductive nor deductive)
Contrapositive: verse? – NOT in Matthew, Mark, Luke

1350. At that day (when Jesus shall shew you plainly of the Father) (then) ye shall ask in my name . . . (Disc) (GAMR) (INFERRED) (Neither inductive nor deductive)
Contrapositive: verse? – NOT in Matthew, Mark, Luke

1351. . . . (since) the Father himself loveth you . . . (then) I (Jesus) say not unto you that I will pray the Father for you. (Disc) (GAGR) (INFERRED) (Neither inductive nor deductive)
Contrapositive: verse? – NOT in Matthew, Mark, Luke

1352. . . . because (since) ye have loved me, and believed that I came out from God, . . . (then) the Father himself loveth you . . . (Disc) (MAGR) (INFERRED) (Neither inductive nor deductive)
Contrapositive: verse? – true vaccuously as the Father loveth all – NOT in Matthew, Mark, Luke

1353. . . . (when) the hour cometh, yea now is come, (then) . . . ye shall be scattered, every man to his own, and shall leave me alone . . . (Disc) (GAMR) (INFERRED) (Neither inductive nor deductive)
Contrapositive: verse? – NOT in Matthew, Mark, Luke

1354. (Since) These things I (Jesus) have spoken unto you (then) . . . in me (Jesus) ye might have peace. (Disc) (GAMR) (INFERRED) (Neither inductive nor deductive)
Contrapositive: verseF? – NOT in Matthew, Mark, Luke

1355. (While) In the world (then) ye shall have tribulation . . . (Disc) (GAMR) (INFEERED) (Deductive)
Contrapositive: verse? – NOT in Matthew, Mark, Luke

1356. . . . (since) I (Jesus) have overcome the world (then) . . . be of good cheer . . . (Disc) (GAMR) (INFERRED) (Deductive)
Contrapositive: verseF? – NOT in Matthew, Mark, Luke

1357. (since the) Father, . . . glorif(ies) the Son, (then) . . . thy Son also may glorify thee . . . (Disc) (GAGR) (INFERRED) (Neither inductive nor deductive)
Contrapositive: verse? – NOT in Matthew, Mark, Luke

1358. (Since) thou (God the Father) hast given him (the Son) power over all flesh, (then) he (the Son) should give eternal life to as many as thou hast given him. (Disc) (GAGR) (INFERRED) (Neither inductive nor deductive)
Contrapositive: verse? – NOT in Matthew, Mark, Luke

1359. . . . now (that I have finished the work that thou gavest me to do), O Father, (then) glorify thou me with thine own self with the glory which I (Jesus) had with thee before the world was. (Disc) (GAGR) (INFERRED) (Neither inductive nor deductive)
Contrapositive: verse? – NOT in Matthew, Mark, Luke

1360. now (that I have finished the work that thou gavest me to do) (then) they (the disciples) have known that all things whatsoever thou hast given me are of thee. (Disc) (GAMR) (INFERRED) (Neither inductive nor deductive)
Contrapositive: verseF? – NOT in Matthew, Mark, Luke

1361. (if) . . . thou (God the Father) hast given me (Jesus) (a thing) (then that thing is) . . . of thee (God the Father). (Disc) (GAGR) (INFERRED) (Deductive)
Contrapositive: verse? – NOT in Matthew, Mark, Luke

1362. . . . (since) I (Jesus) have given unto them the words which thou (God the Father) hast given me, and they (the disciples) have received them, and have surely known that I came out from thee, and they have believed that thou hast sent me (then) . . . they (the disciples) have known that all things whatsoever thou hast given me are of thee . . . (Disc) (GA-MA/MR-GR) (INFERRED) (Inductive since this statement is of the form "If x is a subset of y and if z, then y)
Contrapositive: verse? – NOT in Matthew, Mark, Luke

1363. . . . (since) they (the disciples) are thine . . . (then) I (Jesus) pray not for the world, but for them which thou hast given me . . . (Disc) (GA-MA/GR) (INFERRED) (Neither inductive nor deductive)
Contrapositive: verse? – NOT in Matthew, Mark, Luke

1364. (if a thing is) . . . mine (Jesus') (then that) (is) . . . thine (God the Father's) . . . (Disc) (GAGR) (INFERRED) (Deductive)
Contrapositive: verse? – NOT in Matthew, Mark, Luke

1365. Converse of # 1364—See # 1333 for same – Contrapositive: verse?

1366. . . . now (that I have finished the work that thou gavest me to do) (then) I (Jesus) am no more in the world, but these (the disciples) are in the world, and I come to thee. (Disc) (GA/GR-MR) (INFERRED) (Deductive)
Contrapositive: verse? – NOT in Matthew, Mark, Luke

1367. While I (Jesus) I was with them (the disciples) in the world (then) I kept them in thy name . . . but the son of perdition (Disc) (GAGR) (INFERRED) (Deductive)
Contrapositive: verse? – NOT in Matthew, Mark, Luke

1368. (since) While I (Jesus) was with them (the disciples) in the world (then) I kept them in thy name . . . but the son of perdition (then) . . . the scripture might be fulfilled. (Disc) (GAGR) (INFERRED) (Deductive)
Contrapositive: verse? – NOT in Matthew, Mark, Luke

1369. now (that I have finished the work that thou gavest me to do) (then) come I (Jesus) to thee (God the Father) . . . (Disc) (GAGR) (INFERRED) (Neither inductive nor deductive)
Contrapositive: verse? – very similar to # 1368, but has additional conclusions

1370. (since) . . . these things I (Jesus) speak in the world, (then) . . . they (the disciples) might have my joy fulfilled in themselves. (Disc) (GAMR) (INFERRED) (Deductive)
Contrapositive: verseF? – NOT in Matthew, Mark, Luke

1371. . . . because (since) they (the disciples) are not of the world, even as I (Jesus) am not of the world (then) . . . I have given them thy word; and the world hath hated them . . . (Disc) (MA/GR-MR) (INFERRED) (Deductive)
Contrapositive: verse? – close to # 1303, but additional conclusion

1372. (since) for their sakes I sanctify myself (then) . . . they (the disciples) also might be sanctified through the truth. (Disc) (GA/MR-GR) (INFERRED) (Deductive)
Contrapositive: verse? – true vacuously – NOT in in Matthew, Mark, Luke

1373. (since) . . . pray (not) I (Jesus) for these (the disciples) alone but for them also which shall believe on me through their word, (then) they all may be one: as thou, Father, art in me, and I in thee . . . (Disc) (GAMR) (INFERRED) (Deductive)
Contrapositive: verseF? – NOT in Matthew, Mark, Luke

1374. (since) . . . pray (not) I (Jesus) for these (the disciples) alone but for them also which shall believe on me through their word, (then) they all may be one (then) . . . they also may be one in us . . . (Disc) (GAMR) (INFERRED) (Deductive)
Contrapositive: verseF? – NOT in Matthew, Mark, Luke

1375. (since) . . . pray (not) I (Jesus) for these (the disciples) alone but for them also which shall believe on me through their word, (then) they all may be one (then) . . . they also may be one in us (then) the world may believe that thou hast sent me. (Disc) (GAMR) (INFERRED) (Inductive since this statement is of the form "If x is a subset of y and if z, then y)
Contrapositive: verseF? – NOT in Matthew, Mark, Luke

1376. (since) . . . the glory which thou (God the Father) gavest me I (Jesus) have given them (then) . . . they may be one, even as we are one . . . (Disc) (GAMR) (INFERRED) (Deductive)
Contrapositive: verseF? – NOT in Matthew, Mark, Luke

1377. (since) . . . I (Jesus) (am) . . . in them, and thou (God the Father) in me, (then) they may be made perfect in one; and . . . the world may know that thou hast sent me, and hast loved them, as thou hast loved me. (Disc) (GAMR) (INFERRED) (Deductive)
Contrapositive: verseF? – similar to # 1376, but different initial condition—NOT in Matthew, Mark, Luke

1378. (when) . . . Father, I will that they (the disciples) also, whom thou hast given me, be with me where I am; (then) . . . they may behold my glory, which thou hast given me . . . (Disc) (GAMR) (INFERRED) (Deductive)
Contrapositive: verseF? – NOT in Matthew, Mark, Luke

1379. (since) . . . thou (God the Father) lovedest me before the foundation of the world . . . (then) . . . (when) . . . Father, I will that they (the disciples) also, whom thou hast given me, be with me where I am; (then) . . . they may behold my glory, which thou hast given me . . . (Disc) (GAMR) (INFERRED) (Deductive)
Contrapositive: verseF? – NOT in Matthew, Mark, Luke—see # 1363 for similar, but not same

1380. (since) I (Jesus) have declared unto them (the world) they name, and will declare it (then), the love wherewith thou hast loved me may be in them, and I in them. (Disc) (GAGR) (INFERRED) (Deductive)
Contrapositive: verse? – NOT in Matthew, Mark, Luke

1381. if therefore ye seek me (then) let these go their way. (Disc) (MAMR) (NOT INFERRED) (Neither inductive nor deductive)
Contrapositive: verseF? – NOT in Matthew, Mark, Luke

1382. (since) I (Jesus) have told you that I am he (then) if therefore ye seek me (then) let these go their way. (Disc) (GAMR) (NOT INFERRED) (Neither inductive nor deductive)
Contrapositive: verseF? – NOT in Matthew, Mark, Luke

1383. If I (Jesus) have spoken evil, (then) bear witness of the evil . . . (High priest) (GAMR) (INFERRED- though 'if' given) (Neither inductive nor deductive)
Contrapositive: verseF? – NOT in Matthew, Mark, Luke

1384. . . . if my kingdom were of this world, then would my servants fight, that I should not be delivered to the Jews . . . (Pilate) (GAMR) (NOT INFERRED –'then' given) (Deductive)
Contrapositive: verse? - NOT in Matthew, Mark, Luke

1385. (since) now (that Jesus is before Pilate) (then) is my kingdom not from hence. (Pilate) (GAGR) (INFERRED) (Deductive)
Contrapositive: verse? – NOT in Matthew, Mark, Luke

1386. (since) . . . to this end was I born and for this cause was I born, and for this cause came I into the world (then) I (Jesus) should bear witness unto the truth. (Pilate) (GAGR) (INFERRED) (Neither inductive nor deductive)
Contrapositive: verse? – NOT in Matthew, Mark, Luke

1387. (if) one . . . is of the truth (then that one) heareth my voice. (Pilate) (MAMR) (INFERRED) (Deductive)
Contrapositive: verse? – NOT in Matthew, Mark, Luke

1388. . . . (if) it were (not) given thee from above, (then) . . . Thou couldest have no power at all against me . . . (Pilate) (GAMR) (INFERRED) (Deductive)
Contrapositive: verse? – NOT in Matthew, Mark, Luke

1389. (if) it were (not) given thee from above, (then) . . . Thou couldest have no power at all against me . . . therefore he that hath delivered me unto thee hath the greater sin. (Pilate) (GA-MA/MR) (INFERRED) (Deductive)
Contrapositive: verse? – NOT in Matthew, Mark, Luke

1390. (if one) hath delivered me (Jesus) unto thee (then that one) hath the greater sin. (Pilate) (MAMR) (INFERRED) (Deductive)
Contrapositive: verse? – NOT in Matthew, Mark, Luke

1391. . . . (since) I (Jesus) am not yet ascended to my Father . . . (then) Touch me not . . . . (Mary) (GAMR) (INFERRED) (Neither inductive nor deductive)
Contrapositive: verse? – NOT in Matthew, Mark, Luke

1392. (since) my Father hath sent me (Jesus), (then) even so send I you (the disciples). (Disc) (GAGR) (INFERRED) (Neither inductive nor deductive)
Contrapositive: verse? – NOT in Matthew, Mark, Luke

1393. (if one's) sins ye (the disciples) remit, (then) they are remitted unto them . . . (Disc) (MAGR) (INFERRED) (Neither inductive nor deductive)
Contrapositive: verseF? – NOT in Matthew, Mark, Luke- but see # 161 for similar concept

1394. (if one's) sins ye (the disciples) retain, (then) they are retained. (Disc) (MAGR) (INFERRED) (Neither inductive nor deductive)
Contrapositive: verseF? – NOT in Matthew, Mark, Luke – but see # 162 for similar concept

1395. (since) . . . thou hast seen me, (then) thou hast believed . . . (Thomas) (MA-GA/MR) (INFERRED) (Neither inductive nor deductive)
Contrapositive: verseF? – NOT in Matthew, Mark, Luke – KEY EXCEPTION NOTED for formal human human reasoning—see # 1045

1396. (if one has) . . . not seen, and yet (has) . . . believed (then that one is) . . . blessed . . . (Thomas) (MAGR) (INFERRED) (Neither inductive nor deductive)
Contrapositive: verseF? – NOT in Matthew, Mark, Luke

1397. . . . When thou wast young, (then) thou girdest thyself, and walkedst whither thou wouldest . . . (Peter) (MAMR) (INFERRED) (Neither inductive nor deductive)
Contrapositive: verseF? – NOT in Matthew, Mark Luke

1398. . . . When thou shalt be old, (then) thou shalt stretch forth thy hands, and another shall gird thee, and carry thee whither thou wouldest not . . . (Peter) (MAMR) (INFERRED) (Neither inductive nor deductive)
Contrapositive: verseF? – NOT in Matthew, Mark, or Luke

{
1399. (If a thing is Jesus Christ, then that thing is) . . . the bread of life. (Mult) (GAGR) (INFERRED) (Neither inductive nor deductive)
Contrapositive: verseF? – NOT in Matthew, Mark, Luke – From statement attributed to Christ, ". . . I am the bread of life . . ."

(If a thing is) . . . the bread of life, (then that thing is) . . . I (Jesus Christ). (Mult) (GAGR) (INFERRED) (Neither inductive nor deductive)
Contrapositive: verseF? – NOT in Matthew, Mark, Luke – From statement attributed to Christ, ". . . I am the bread of life . . ." (which also results in converse holding since this is a definition
}

{
1400. (If a thing is Jesus Christ, then that thing is) . . . that bread of life. (Mult) (GAGR) (INFERRED) (Neither inductive nor deductive)
Contrapositive: verseF? – NOT in Matthew, Mark, Luke – From statement attributed to Christ, ". . . I am that bread of life . . ."

(If a thing is) . . . that bread of life, (then that thing is) . . . I (Jesus Christ). (Mult) (GAGR) (INFERRED) (Neither inductive nor deductive)
Contrapositive: verseF? – NOT in Matthew, Mark, Luke – From statement attributed to Christ, ". . . I am that bread of life . . ." (which also results in converse holding since this is a definition
}
{
1401. (If a thing is Jesus Christ, then that thing is) . . . the living bread that came down from heaven. . . . (Mult) (GAGR) (INFERRED) (Neither inductive nor deductive)
Contrapositive: verseF? – NOT in Matthew, Mark, Luke – From statement attributed to Christ, ". . . I am the living bread that came down from heaven . . ."

(If a thing is) . . . the living bread that came down from heaven, (then that thing is) . . . I (Jesus Christ). (Mult) (GAGR) (INFERRED) (Neither inductive nor deductive)
Contrapositive: verseF? – NOT in Matthew, Mark, Luke – From statement attributed to Christ, ". . . I am the living bread that came down from heaven . . ." (which also results in converse holding since this is a definition
}
{

1402. (If a thing is Jesus Christ, then that thing is) . . . the light of the world. (Mult/Relig/Disc) (GAGR) (INFERRED) (Neither inductive nor deductive)
Contrapositive: verseF? – NOT in Matthew, Mark, Luke – From statement attributed to Christ, ". . . I am the light of the world . . ."

The converse holds given # 1495, # 1101, and #1103 in which Christ is attributed as saying, "Ye are the light of the world."
}
{
1403. (If a thing is Jesus Christ, then that thing is) . . . the door of the sheep. (Mult) (GAGR) (INFERRED) (Neither inductive nor deductive)
Contrapositive: verseF? – NOT in Matthew, Mark, Luke – From statement attributed to Christ, ". . . I am the door of the sheep . . ."

(If a thing is) . . . the door of the sheep, (then that thing is) . . . I (Jesus Christ). (Mult) (GAGR) (INFERRED) (Neither inductive nor deductive)
Contrapositive: verseF? – NOT in Matthew, Mark, Luke – From statement attributed to Christ, ". . . I am the door of the sheep . . ." (which also results in converse holding since this is a definition
}
{
1404. (If a thing is Jesus Christ, then that thing is) . . . the good shepherd. (Mult) (GAGR) (INFERRED) (Neither inductive nor deductive)
Contrapositive: verseF? – NOT in Matthew, Mark, Luke – From statement attributed to Christ, ". . . I am the good shepherd . . ."

(If a thing is) . . . the good shepherd, (then that thing is) . . . I (Jesus Christ). (Mult) (GAGR) (INFERRED) (Neither inductive nor deductive)
Contrapositive: verseF? – NOT in Matthew, Mark, Luke – From statement attributed to Christ, ". . . I am the good shepherd . . ." (which also results in converse holding since this is a definition
}
{
1405 (If a thing is Jesus Christ, then that thing is) . . . one (with)(God) the Father. (Mult/Relig/Disc) (GAGR) (INFERRED) (Neither inductive nor deductive)
Contrapositive: verseF? – NOT in Matthew, Mark, Luke – From statement attributed to Christ, ". . . I and the Father are one . . ." (see # 1272 for why contrapositive would also be false)

(If a thing is) . . . (God) the Father, (then that thing is) . . . (Jesus Christ). (Mult/Relig/Disc) (GAGR) (INFERRED) (Neither inductive nor deductive)

Contrapositive: verseF? – NOT in Matthew, Mark, Luke – From statement attributed to Christ, ". . . I and the Father are one . . ." (which also results in converse holding since this is a definition (see # 1272 for why contrapositive would also be false)

}
{
1406. (If a thing is Jesus Christ, then that thing is) . . . the Son of God . . . . (Mult/Relig/Disc) (GAGR) (INFERRED) (Neither inductive nor deductive)
Contrapositive: verseF? – NOT in Matthew, Mark, Luke – From statement attributed to Christ, ". . . I am the Son of God . . ."

(If a thing is) . . . the Son of God, (then that thing is) . . . I (Jesus Christ). (Mult/Relig/Disc) (GAGR) (INFERRED) (Neither inductive nor deductive)
Contrapositive: verseF? – NOT in Matthew, Mark, Luke – From statement attributed to Christ, ". . . I am the Son of God . . ." (which also results in converse holding since this is a definition
}
{
1407. (If a thing is Jesus Christ, then that thing is) . . . the resurrection and the life . . . (Mult) (GAGR) (INFERRED) (Neither inductive nor deductive)
Contrapositive: verseF? – NOT in Matthew, Mark, Luke – From statement attributed to Christ, ". . . I am the resurrection and the life . . ."

(If a thing is) . . . the resurrection and the life, (then that thing is) . . . I (Jesus Christ). (Mult) (GAGR) (INFERRED) (Neither inductive nor deductive)
Contrapositive: verseF? – NOT in Matthew, Mark, Luke – From statement attributed to Christ, ". . . I am the resurrection and the life . . ." (which also results in converse holding since this is a definition
}
{
1408. (If a thing is Jesus Christ, then that thing is) . . . the Master and Lord. (Disc) (GAGR) (INFERRED) (Neither inductive nor deductive)
Contrapositive: verseF? – NOT in Matthew, Mark, Luke – From statement attributed to Christ, ". . . I am (Master and Lord) . . ."

(If a thing is) . . . Master and Lord, (then that thing is) . . . I (Jesus Christ). (Disc) (GAGR) (INFERRED) (Neither inductive nor deductive)
Contrapositive: verseF? – NOT in Matthew, Mark, Luke – From statement attributed to Christ, ". . . I am (Master and Lord) . . ." (which also results in converse holding since this is a definition
}
{

1409. (If a thing is Jesus Christ, then that thing is) . . . the way . . . . (Disc) (GAGR) (INFERRED) (Neither inductive nor deductive)
Contrapositive: verseF? – NOT in Matthew, Mark, Luke – From statement attributed to Christ, ". . . I am the way . . ."

(If a thing is) . . . the way, (then that thing is) . . . I (Jesus Christ). (Disc) (GAGR) (INFERRED) (Neither inductive nor deductive)
Contrapositive: verseF? – NOT in Matthew, Mark, Luke – From statement attributed to Christ, ". . . I am the way . . ." (which also results in converse holding since this is a definition
}
{
1410. (If a thing is Jesus Christ, then that thing is) . . . the truth. . . . (Disc) (GAGR) (INFERRED) (Neither inductive nor deductive)
Contrapositive: verseF? – NOT in Matthew, Mark, Luke – From statement attributed to Christ, ". . . I am . . . the truth . . ."

(If a thing is) . . . the truth . . . , (then that thing is) . . . I (Jesus Christ). (Disc) (GAGR) (INFERRED) (Neither inductive nor deductive)
Contrapositive: verseF? – NOT in Matthew, Mark, Luke – From statement attributed to Christ, ". . . I am the truth . . ." (which also results in converse holding since this is a definition
}
{
1411. (If a thing is Jesus Christ, then that thing is) . . . the true vine . . . (Disc) (GAGR) (INFERRED) (Neither inductive nor deductive)
Contrapositive: verseF? – NOT in Matthew, Mark, Luke – From statement attributed to Christ, ". . . I am the true vine . . ."

(If a thing is) . . . the true vine . . . , (then that thing is) . . . I (Jesus Christ). (Disc) (GAGR) (INFERRED) (Neither inductive nor deductive)
Contrapositive: verseF? – NOT in Matthew, Mark, Luke – From statement attributed to Christ, ". . . I am the true vine . . ." (which also results in converse holding since this is a definition
}
{
1412. (If a human is Jesus Christ, then that thing is) . . . (Jesus of Nazareth) . . . (Disc) (GAGR) (INFERRED) (Neither inductive nor deductive)
Contrapositive: verseF? – NOT in Matthew, Mark, or Luke – From statement attributed to Christ, ". . . I am he (Jesus of Nazareth) . . ."

(If a human is) . . . (Jesus of Nazareth) . . . , (then that human is) . . . I (Jesus Christ). (Disc) (GAGR) (INFERRED) (Neither inductive nor deductive)
Contrapositive: verseF? – NOT in Matthew, Mark, or Luke – From statement attributed to Christ, ". . . I am he (Jesus of Nazareth) . . ." (which also results in converse holding since this is a definition
}
{
1413. (If a human is Simon the Apostle, then that human is) . . . Peter (the rock upon which Christ would build his church) . . . (Simon) (GAGR) (INFERRED) (Neither inductive nor deductive)
Contrapositive: verseF? – NOT in Mark, Luke, or John - From statement attributed to Christ, ". . . Simon . . . thou art Peter . . ."

(If a human is) . . . Peter (the rock upon which Christ would build his church) . . . , (then that human is Simon the Apostle). (Simon) (GAGR) (INFERRED) (Neither inductive nor deductive)
Contrapositive: verseF? – NOT in Mark, Luke, or John – From statement attributed to Christ, ". . . Simon . . . thou art Peter . . ." (which also results in converse holding since this is a definition
}
{
1414. (If a human is Simon the Apostle, then that human is) . . . Cephas (the stone) . . . (Simon) (GAGR) (INFERRED) (Neither inductive nor deductive)

Contrapositive: verseF? – NOT in Matthew, Mark, or Luke - From statement attributed to Christ, ". . . Simon . . . thou art Cephas . . ."
(If a human is) . . . Cephas (the stone) . . . , (then that human is Simon the Apostle). (Simon) (GAGR) (INFERRED) (Neither inductive nor deductive)
Contrapositive: verseF? – NOT in Matthew, Mark, or Luke – From statement attributed to Christ, ". . . Simon . . . thou art Cephas . . ." (which also results in converse holding since this is a definition
}
1415. (If a thing is heaven, then that thing is) . . . God's Throne . . . (Mult) (GAGR) (INFERRED) (Neither inductive nor deductive)
Contrapositive: verseF? – NOT in Mark, Luke, or John – From statement attributed to Christ, ". . . heaven, for it is God's Throne . . ."

(If a thing is) . . . God's Throne . . . , (then that thing is) . . . heaven . . . (Mult) (GAGR) (INFERRED) (Deductive)

Contrapositive: verseF? – NOT in Mark, Luke, or John – From statement attributed to Christ, ". . . heaven, for it is God's Throne . . ." (which also results in converse holding since this is a definition
}
{
1416. (If a thing is the earth, then that thing is) . . . (God's) footstool . . . (Mult) (GAGR) (INFERRED) (Neither inductive nor deductive)
Contrapositive: verseF? – NOT in Mark, Luke, or John – From statement attributed to Christ, ". . . earth, for it is his (God's) footstool . . ."

(If a thing is) . . . (God's footstool) . . . , (then that thing is) . . . the earth . . . (Mult) (GAGR) (INFERRED) (Neither inductive nor deductive)
Contrapositive: verseF? – NOT in Mark, Luke, or John – From statement attributed to Christ, ". . . earth, for it is his (God's) footstool . . ." (which also results in converse holding since this is a definition
}
{
1417. (If a thing is Jerusalem, then that thing is) . . . the city of the great King . . . (Mult) (GAGR) (INFERRED) (Neither inductive nor deductive)
Contrapositive: verseF? – NOT in Mark, Luke, or Luke – From statement attributed to Christ, ". . . Jerusalem, for it is the city of the great King . . ."

(If a thing is) . . . the city of the great King . . . , (then that thing is) . . . Jerusalem . . . (Mult) (GAGR) (INFERRED) (Neither inductive nor deductive)
Contrapositive: verseF? – NOT in Mark, Luke, or John – From statement attributed to Christ, ". . . Jerusalem, for it is the city of the great King . . ." (which also results in converse holding since this is a definition
}
{
1418. (If a thing is light of the body, then that thing is) . . . the eye (of a human) . . . (Mult) (GAGR) (INFERRED) (Deductive)
Contrapositive: verseF? – NOT in Mark, Luke, or John – From statement attributed to Christ, "The light of the body is the eye . . ."

(If a thing is) . . . the eye (of a human) . . . , (then that thing is) . . . the light of the body. (Mult) (GAGR) (INFERRED) (Deductive)
Contrapositive: verseF? – NOT in Mark, Luke, or John – From statement attributed to Christ, "The light of the body is the eye . . ." (which also results in converse holding since this is a definition
}
{

1419. (If a thing is such that) whatsoever ye would that men should do to you, do ye even so to them . . . (then that thing is 'the general summary of the teachings' in) . . . the law and the prophets. (Mult) (GAGR) (INFERRED) (Deductive)
Contrapositive: verseF? – NOT in Mark, Luke, or John – From statement attributed to Christ, "Therefore all things whatsoever ye would that men should do to you, do ye even so to them: for this is the law and the prophets."

(If a thing is 'the general summary of the teachings' in) . . . the law and the prophets) . . . (then that thing is such that) whatsoever ye would that men should do to you, do ye even so to them . . . (Mult) (GAGR) (INFERRED) (Deductive)
Contrapositive: verseF? – NOT in Mark, Luke, or John – From statement attributed to Christ, "Therefore all things whatsoever ye would that men should do to you, do ye even so to them: for this is the law and the prophets." (which also results in converse holding since this is a definition
}
{
1420. (If a thing is him (Satan), then that thing is that) . . . which is able to destroy both soul and body in hell . . . (Disc) (GAGR) (INFERRED) (Neither inductive nor deductive)
Contrapositive: verseF? – NOT in Mark, Luke, or John – From statement attributed to Christ, ". . . him which is able to destroy both soul and body in hell."

(If a thing is) . . . able to destroy both soul and body in hell . . . , (then that thing is) . . . him (Satan). (Disc) (GAGR) (INFERRED) (Neither inductive nor deductive)
Contrapositive: verseF? – NOT in Mark, Luke, or John – From statement attributed to Christ, ". . . him which is able to destroy both soul and body in hell." (which also results in converse holding since this is a definition
}
{
1421. (If a human is this (John the Baptist), then that human is) . . . Elias which was for to come . . . (Disc) (GAGR) (INFERRED) (Neither inductive nor deductive)
Contrapositive: verseF? – NOT in Mark, Luke, or John – From statement attributed to Christ, ". . . this is Elias, which was for to come."

(If a human is) . . . Elias, which was for to come (,then that human is) . . . this (John the Baptist). (Disc) (GAGR) (INFERRED) (Neither inductive nor deductive)

Contrapositive: verseF? – NOT in Mark, Luke, or John – From statement attributed to Christ, ". . . this is Elias, which was for to come" (which also results in converse holding since this is a definition
}
{
1422. (If a human is a) . . . disciple (,then that human is) . . . not above his master . . . (Disc) (GAGR) (INFERRED) (Deductive)
Contrapositive: verseF? – NOT in Mark, Luke, or John – From statement attributed to Christ, "The disciple is not above his master . . ."

(If a human is) . . . not above his master . . . , (then that human is a) . . . disciple. (Disc) (GAGR) (INFERRED) (Deductive)
Contrapositive: verseF? – NOT in Mark, Luke, or John – From statement attributed to Christ, ". . . I am the true vine . . ." (which also results in converse holding since this is a definition
}
{
1423. (If a human is a) . . . servant (,then that human is not) . . . above his lord . . . (Disc) (GAGR) (INFERRED) (Deductive)
Contrapositive: verseF? – NOT in Mark, Luke, or John – From statement attributed to Christ, ". . . nor the servant above his lord."

(If a human is not) . . . above his lord . . . , (then that human is a) . . . servant. (Disc) (GAGR) (INFERRED) (Deductive)
Contrapositive: verseF? – NOT in Mark, Luke, or John – From statement attributed to Christ, ". . . nor the servant above his lord." (which also results in converse holding since this is a definition
}
{
1424. (If a thing is him (Satan), then that thing is that) . . . which is able to destroy both soul and body in hell . . . (Disc) (GAGR) (INFERRED) (Neither inductive nor deductive)
Contrapositive: verseF? – NOT in Mark, Luke, or John – From statement attributed to Christ, ". . . him which is able to destroy both soul and body in hell."

(If a thing is) . . . able to destroy both soul and body in hell . . . , (then that thing is) . . . him (Satan). (Disc) (GAGR) (INFERRED) (Neither inductive nor deductive)
Contrapositive: verseF? – NOT in Mark, Luke, or John – From statement attributed to Christ, ". . . I am the true vine . . ." (which also results in converse holding since this is a definition

}
{
1425. (If a human is) . . . the Son of man . . . , (then that human is) . . . Lord even of the sabbath day . . . (Disc) (GAGR) (INFERRED) (Neither inductive nor deductive)
Contrapositive: verseF? – NOT in Mark, Luke, or John – From statement attributed to Christ, ". . . the Son of man is Lord even of the sabbath day."

(If a human is) . . . Lord even of the sabbath day, (then that human is) . . . the Son of man. (Disc) (GAGR) (INFERRED) (Neither inductive nor deductive)
Contrapositive: verseF? – NOT in Mark, Luke, or John – From statement attributed to Christ, ". . . the Son of man is Lord even of the sabbath day." (which also results in converse holding since this is a definition
}
{
1426. (If a human) . . . heareth the word of the kingdom, and understandeth it not, then cometh the wicked one, and catcheth away that which was sown in his heart . . . (,then that human is that) . . . which received seed by the way side. (Disc) (GA-MA/GR) (INFERRED) (Deductive)
Contrapositive: verseF? – From statement attributed to Christ, ". . . any one heareth the word of the kingdom, and understandeth it not, then cometh the wicked one, and catcheth away that which was sown in his heart. This is he which received seed by the way side."

(If a human is that which) . . . which received seed by the way side. . . . , (then that human is) . . . one (that) heareth the word of the kingdom, and understandeth it not, then cometh the wicked one, and catcheth away that which was sown in his heart . . . (Disc) (GA-MA/GR) (INFERRED) (Deductive)
Contrapositive: verseF? – – From statement attributed to Christ, ". . . any one heareth the word of the kingdom, and understandeth it not, then cometh the wicked one, and catcheth away that which was sown in his heart. This is he which received seed by the way side." (which also results in converse holding since this is a definition
}
{
1427. (If a human is that which) . . . received the seed into stony places . . . (,then that human is that which) . . . heareth the word, and anon with joy receiveth it;Yet hath he not root in himself, but dureth for a while: for when tribulation or persecution ariseth because of the word, by and by he is offended. (Disc) (GA-MA/GR) (INFERRED) (Deductive)

Contrapositive: verseF? – From statement attributed to Christ, "... he that received the seed into stony places, the same is he that heareth the word, and anon with joy receiveth it; Yet hath he not root in himself, but dureth for a while: for when tribulation or persecution ariseth because of the word, by and by he is offended."

(If a human is that which) . . . heareth the word, and anon with joy receiveth it; Yet hath he not root in himself, but dureth for a while: for when tribulation or persecution ariseth because of the word, by and by he is offended.. . . . , (then that human is) . . . one (that) received the seed into stony places . . . (Disc) (GA-MA/GR) (INFERRED) (Deductive)
Contrapositive: verseF? – – From statement attributed to Christ, ". . . he that received the seed into stony places, the same is he that heareth the word, and anon with joy receiveth it; Yet hath he not root in himself, but dureth for a while: for when tribulation or persecution ariseth because of the word, by and by he is offended." (which also results in converse holding since this is a definition
}
{
1428. (If a human) . . . heareth the word of the kingdom, and understandeth it not, then cometh the wicked one, and catcheth away that which was sown in his heart . . . (,then that human is that) . . . which received seed by the way side . . . (Disc) (GA-MA/GR) (INFERRED) (Deductive)
Contrapositive: verseF? – From statement attributed to Christ, ". . . any one heareth the word of the kingdom, and understandeth it not, then cometh the wicked one, and catcheth away that which was sown in his heart. This is he which received seed by the way side."

(If a human is that which) . . . which received seed among the thorns. . . . , (then that human is) . . . one (that) . . . heareth the word; and the care of this world, and the deceitfulness of riches, choke the word, and he becometh unfruitful. (Disc) (GA-MA/GR) (INFERRED) (Deductive)
Contrapositive: verseF? – From statement attributed to Christ, ". . . he that received seed among the thorns is he that heareth the word; and the care of this world, and the deceitfulness of riches, choke the word, and he becometh unfruitful." (which also results in converse holding since this is a definition
}
{
1429. (If a human is that which) . . . heareth the word; and the care of this world, and the deceitfulness of riches, choke the word, and he becometh un-

fruitful. . . . (,then that human is that which) . . . received seed among the thorns . . . (Disc) (GA-MA/GR) (INFERRED) (Deductive)
Contrapositive: verseF? – From statement attributed to Christ, ". . . he that received seed among the thorns is he that heareth the word; and the care of this world, and the deceitfulness of riches, choke the word, and he becometh unfruitful."

(If a human is that which) . . . received seed by the way side. . . . , (then that human is) . . . one (that) heareth the word of the kingdom, and understandeth it not, then cometh the wicked one, and catcheth away that which was sown in his heart . . . (Disc) (GA-MA/GR) (INFERRED) (Deductive)
Contrapositive: verseF? – – From statement attributed to Christ, ". . . any one heareth the word of the kingdom, and understandeth it not, then cometh the wicked one, and catcheth away that which was sown in his heart. This is he which received seed by the way side." (which also results in converse holding since this is a definition
}
{
1430. (If a human is that which) . . . received seed into the good ground . . . (,then that human) . . . is he that heareth the word, and understandeth it; which also beareth fruit, and bringeth forth, some an hundredfold, some sixty, some thirty. . . . (Disc) (GA-MA/GR) (INFERRED) (Deductive)
Contrapositive: verseF? – From statement attributed to Christ, ". . . he that received seed into the good ground is he that heareth the word, and understandeth it; which also beareth fruit, and bringeth forth, some an hundredfold, some sixty, some thirty."

(If a human) . . . is he that heareth the word, and understandeth it; which also beareth fruit, and bringeth forth, some an hundredfold, some sixty, some thirty . . . , (then that human is that which) . . . received seed into the good ground . . . (Disc) (GA-MA/GR) (INFERRED) (Deductive)
Contrapositive: verseF? – From statement attributed to Christ, ". . . any one heareth the word of the kingdom, and understandeth it not, then cometh the wicked one, and catcheth away that which was sown in his heart. This is he which received seed by the way side." (which also results in converse holding since this is a definition)
}
{
1431. (If a thing is) . . . a grain of mustard seed . . . (,then that thing is) . . . the least of all seeds: but when it (has been planted in the field and) is grown, it is the greatest among herbs . . . (Disc) (GAGR) (INFERRED) (Deductive)

Contrapositive: verseF? - From statement attributed to Christ, ". . . a grain of mustard seed, which a man took, and sowed in his field: Which indeed is the least of all seeds: but when it is grown, it is the greatest among herbs . . ."

(If a thing is) . . . the least of all seeds: but when it (has been planted in the field and) is grown, it is the greatest among herbs . . . (,then that thing is) . . . a grain of mustard seed . . . (Disc) (GAGR) (INFERRED) (Deductive)
Contrapositive: verseF? - From statement attributed to Christ, ". . . a grain of mustard seed, which a man took, and sowed in his field: Which indeed is the least of all seeds: but when it is grown, it is the greatest among herbs . . ." (which also results in converse holding since this is a definition)
}
{
1432. (If a human is) . . . He that soweth the good seed . . . (,then that human is) . . . the Son of man . . . (Disc) (GAGR) (INFERRED) (Neither inductive nor deductive)
Contrapositive: verseF? - From statement attributed to Christ, ". . . He that soweth the good seed is the Son of man;"

(If a human is) . . . the Son of man . . . (,then that human is) . . . He that soweth the good seed . . . (Disc) (GAGR) (INFERRED) (Neither inductive nor deductive)
Contrapositive: verseF? - From statement attributed to Christ, ". . . He that soweth the good seed is the Son of man . . ." (which also results in converse holding since this is a definition)
}
{
1433. (If a thing is) . . . the field (in which the seed is sown) . . . (,then that thing is) . . . the world . . . (Disc) (GAGR) (INFERRED) (Neither inductive nor deductive)
Contrapositive: verseF? - From statement attributed to Christ, ". . . the field is the world . . ."

(If a thing is) . . . the world . . . (,then that thing is) . . . the field (into which the seed is sown) . . . (Disc) (GAGR) (INFERRED) (Neither inductive nor deductive)
Contrapositive: verseF? - From statement attributed to Christ, ". . . the field is the world . . ." (which also results in converse holding since this is a definition)
}
{

1434. (If a thing is) . . . the good seed . . . (,then that thing is) . . . the children of the kingdom . . . (Disc) (GAGR) (INFERRED) (Neither inductive nor deductive)
Contrapositive: verseF? - From statement attributed to Christ, ". . . the good seed are the children of the kingdom . . ."

(If a thing is) . . . the children of the kingdom . . . (,then that thing is) . . . the good seed . . . (Disc) (GAGR) (INFERRED) (Neither inductive nor deductive)
Contrapositive: verseF? - From statement attributed to Christ, ". . . the good seed are the children of the kingdom . . ." (which also results in converse holding since this is a definition)
}
{
1435. (If a thing is a) . . . tare . . . (,then that thing is a) . . . child . . . of the wicked one . . . (Disc) (GAGR) (INFERRED) (Neither inductive nor deductive)
Contrapositive: verseF? - From statement attributed to Christ, ". . . the tares are the children of the wicked one . . ."

(If a thing is a) . . . child of the wicked one . . . (,then that thing is a) . . . tare . . . (Disc) (GAGR) (INFERRED) (Neither inductive nor deductive)
Contrapositive: verseF? - From statement attributed to Christ, ". . . the tares are the children of the wicked one . . ."(which also results in converse holding since this is a definition)
}
{
1436. (If a thing is) . . . the enemy that sowed the tares . . . (,then that thing is) . . . the devil . . . (Disc) (GAGR) (INFERRED) (Neither inductive nor deductive)
Contrapositive: verseF? - From statement attributed to Christ, ". . . the enemy that sowed the tares is the devil . . ."

(If a thing is) . . . the devil . . . (,then that thing is) . . . the enemy that sowed the tares . . . (Disc) (GAGR) (INFERRED) (Neither inductive nor deductive)
Contrapositive: verseF? - From statement attributed to Christ, ". . . the enemy that sowed the tares is the devil . . ." (which also results in converse holding since this is a definition)
}
{

1437. (If a thing is) . . . the harvest (of the seeds and tares in the field) . . . (,then that thing is) . . . the end of the world . . . (Disc) (GAGR) (INFERRED) (Neither inductive nor deductive)
Contrapositive: verseF? - From statement attributed to Christ, ". . . the harvest is the end of the world . . ."

(If a thing is) . . . the end of the world . . . (,then that thing is) . . . the harvest (of the seeds and tares in the field) . . . (Disc) (GAGR) (INFERRED) (Neither inductive nor deductive)
Contrapositive: verseF? - From statement attributed to Christ, ". . . the harvest is the end of the world . . ." (which also results in converse holding since this is a definition)
}
{
1438. (If a thing is a) . . . reaper (of the seeds and tares in the field) . . . (,then that thing is an) . . . angel . . . (Disc) (GAGR) (INFERRED) (Neither inductive nor deductive)
Contrapositive: verseF? - From statement attributed to Christ, ". . . the reapers are the angels . . ."

(If a thing is an) . . . angel . . . (,then that thing is a) . . . reaper (of the seeds and tares in the field) . . . (Disc) (GAGR) (INFERRED) (Neither inductive nor deductive)
Contrapositive: verseF? - From statement attributed to Christ, ". . . the reapers are the angels . . ." (which also results in converse holding since this is a definition)
}
{
1439. (If a thing is) . . . the will of your Father in heaven . . . (,then that thing is such that) . . . one of these little ones should (not) perish . . . (Disc) (GAGR) (INFERRED) (Neither inductive nor deductive) - NOT in Mark, Luke, John
Contrapositive: verseF? - From statement attributed to Christ, ". . . Even so it is not the will of your Father which is in heaven, that one of these little ones should perish. . . ."

(If a thing is such that) . . . one of these little ones should perish . . . (,then that thing is) . . . not the will of your Father in heaven . . . (Disc) (GAGR) (INFERRED) (Neither inductive nor deductive) - NOT in Mark, Luke, John
Contrapositive: verseF? - From statement attributed to Christ, ". . . Even so it is not the will of your Father which is in heaven, that one of these little ones

should perish. . . ." (which also results in converse holding since this is a definition)
}
{
1440. (If a thing can be done) . . . (,then that thing is) . . . with God . . . possible . . . (Disc) (GAGR) (INFERRED) (Neither inductive nor deductive) - NOT in Mark, Luke, John
Contrapositive: verseF? - From statement attributed to Christ, ". . . with God all things are possible . . ."

(If a thing is such that it is) . . . possible . . . with God . . . (,then that thing exists) . . . (Disc) (GAGR) (INFERRED) (Neither inductive nor deductive) - NOT in Mark, Luke, John
Contrapositive: verseF? - From statement attributed to Christ, ". . . with God all things are possible. . . ." (which also results in converse holding since this is a definition)
}
{
1441. (If a commandment is such that) . . . Thou shalt love the Lord thy God with all thy heart, and with all thy soul, and with all thy mind . . . (then that commandment) . . . is the first and great commandment (Relig) (GAGR) (INFERRED) (Deductive) - NOT in Mark, Luke, John
Contrapositive: verseF? – From statement attributed to Christ as, "Thou shalt love the Lord thy God with all thy heart, and with all thy soul, and with all thy mind. This is the first and great commandment."

(If a commandment is such that it) . . . is the first and great commandment . . . (,then that commandment is such that) . . . Thou shalt love the Lord thy God with all thy heart, and with all thy soul, and with all thy mind . . . (Relig) (GAGR) (INFERRED) (Deductive) - NOT in Mark, Luke, John Contrapositive: verseF? – From statement attributed to Christ as, "Thou shalt love the Lord thy God with all thy heart, and with all thy soul, and with all thy mind. This is the first and great commandment." (which also results in converse holding since this is a definition)
}
{
1442. (If a commandment is such that) . . . Thou shalt love thy neighbor as thyself . . . (then that commandment) . . . is the second (greatest) commandment (Relig) (GAGR) (INFERRED) (Deductive) - NOT in Mark, Luke, John
Contrapositive: verseF? – From statement attributed to Christ as, "And the second is like unto it, Thou shalt love thy neighbour as thyself."

(If a commandment is such that it) . . . is the second (greatest) commandment . . . (,then that commandment is such that) . . . Thou shalt love thy neighbor as thyself . . . (Relig) (GAGR) (INFERRED) (Deductive) - NOT in Mark, Luke, John
Contrapositive: verseF? – From statement attributed to Christ as, "And the second is like unto it, Thou shalt love thy neighbour as thyself." (which also results in converse holding since this is a definition)
}
{
1443. (If a human is) . . . greatest among you . . . (,then that human) shall be your servant. (Disc) (GA-MA/MR) (INFERRED) (Inductive) - NOT in Mark, Luke, John
Contrapositive: verseF? – From statement attributed to Christ as, "But he that is greatest among you shall be your servant."

(If a human) . . . shall be your servant (,then that human) . . . is greatest among you . . . (Disc) (MA/GR-MR) (INFERRED) (Deductive) - NOT in Mark, Luke, John
Contrapositive: verseF? – From statement attributed to Christ as, "But he that is greatest among you shall be your servant." (which also results in converse holding since this is a definition)
}
{
1444. (If a thing is such that it is) . . . the end (of time) . . . (, then that thing) . . . is not yet (where 'yet' is defined as when the disciples of Christ "shall hear of wars and rumours of wars"). (Disc) (GA/GR-MR) (INFERRED) (Deductive) - NOT in Mark, Luke, John
Contrapositive: verseF? – From statement attributed to Christ as, "And ye shall hear of wars and rumours of wars: see that ye be not troubled: for all these things must come to pass, but the end is not yet."

(If a thing is such that it) . . . is not yet (where 'yet' is defined as when the disciples of Christ "shall hear of wars and rumours of wars") . . . (, then that thing is) . . . the end (of time). (Disc) (GA-MA/GR) (INFERRED) (Deductive) - NOT in Mark, Luke, John
Contrapositive: verseF? – From statement attributed to Christ as, "And ye shall hear of wars and rumours of wars: see that ye be not troubled: for all these things must come to pass, but the end is not yet." (which also results in converse holding since this is a definition)
}

1445. (If a thing is) . . . this (the bread used during Christ's 'Last Supper' with the disciples) . . . (, then that thing) . . . is my (Christ's) body. (Disc) (GAGR) (INFERRED) (Neither inductive nor deductive) - NOT in Mark, Luke, John
Contrapositive: verseF? – From statement attributed to Christ as, ". . . this is my body." (1st KEY NOTE: The converse of this statement cannot necessarily assumed to be TRUE or FALSE based solely on the words attributed to Christ in the Holy Bible since Christ stated in Luke 22: 16, 18 that, "I (Jesus Christ) will not eat/drink (of the bread or "the fruit of the vine") until it be fulfilled in the kingdom of God/the kingdom of God shall come." {2nd KEY NOTE: This passage is included in this analysis because an over-arching assumption of this book is that in order to compare Christ's use of Logic to the formal rules of human reasoning requires strictly use of literal (i.e., no symbolic or figurative interpretations), and the word "is" in this context is the same as all other instances of 'definitions' included in this analysis. Moreover, such inclusion here does not remove the possibility of a symbolic or figurative interpretation for this passage, but rather restricts the attention for the purposes of this analysis to considerations limited solely to literal meanings.

1446. (If a thing is) . . . this (the drink used during Christ's 'Last Supper' with the disciples) . . . (, then that thing) . . . is my (Christ's) blood of the new testament. (Disc) (GAGR) (INFERRED) (Neither inductive nor deductive) - NOT in Mark, Luke, John
Contrapositive: verseF? – From statement attributed to Christ as, ". . . this is my blood of the new testament . . ." (1st KEY NOTE: The converse of this statement cannot necessarily assumed to be TRUE or FALSE based solely on the words attributed to Christ in the Holy Bible since Christ stated in Luke 22: 16, 18 that, "I (Jesus Christ) will not eat/drink (of the bread or "the fruit of the vine") until it be fulfilled in the kingdom of God/the kingdom of God shall come."{2nd KEY NOTE: This passage is included in this analysis because an over-arching assumption of this book is that in order to compare Christ's use of Logic to the formal rules of human reasoning requires strictly use of literal (i.e., no symbolic or figurative interpretations), and the word "is" in this context is the same as all other instances of 'definitions' included in this analysis. Moreover, such inclusion here does not remove the possibility of a symbolic or figurative interpretation for this passage, but rather restricts the attention for the purposes of this analysis to considerations limited solely to literal meanings.

{
1447. (If a human is) . . . this (the son of the certain man (who) planted a vineyard) . . . (then that human is) . . . the heir (to the certain man (who) planted a vineyard) . . . (Disc) (GAGR) (INFERRED) (Neither inductive nor deductive) - NOT in Mark, Luke, John
Contrapositive: verseF? – From statement attributed to Christ as, ". . . This is the heir . . ."

(If a human is) . . . the heir (to the certain man (who) planted a vineyard) . . . (then that human is) . . . this (the son of the certain man (who) planted a vineyard) . . . (Disc) (GAGR) (INFERRED) (Neither inductive nor deductive) - NOT in Mark, Luke, John
Contrapositive: verseF? – From statement attributed to Christ as, ". . . This is the heir . . ." (which also results in converse holding since this is a definition)
}

{
1448. (If a thing is) . . . He (the God of Abraham, the God of Isaac, and the God of Jacob) . . . (, then that thing) . . . is not the God of the dead, but the God of living. (Disc) (GAGR) (INFERRED) (Neither inductive nor deductive) - NOT in Mark, Luke, John
Contrapositive: verseF? – From statement attributed to Christ as, "He is not the God of the dead, but the God of living . . ."

(If a thing) . . . is not the God of the dead, but the God of living (, then that human is) . . . He (the God of Abraham, the God of Isaac, and the God of Jacob) . . . (Disc) (GAGR) (INFERRED) (Neither inductive nor deductive) - NOT in Mark, Luke, John
Contrapositive: verseF? – From statement attributed to Christ as, ". . . this is my blood of the new testament." (which also results in converse holding since this is a definition)
}
{
1449. (If a thing is) . . . the Lord our God . . . (, then that thing) . . . is one Lord. (Disc) (GAGR) (INFERRED) (Neither inductive nor deductive) - NOT in Mark, Luke, John
Contrapositive: verseF? – From statement attributed to Christ as, "The Lord our God is one Lord . . ."

(If a thing is) . . . one Lord . . . (, then that thing is) . . . the Lord our God. (Disc) (GAGR) (INFERRED) (Neither inductive nor deductive) - NOT in Mark, Luke, John

Contrapositive: verseF? – From statement attributed to Christ as, "The Lord our God is one Lord . . ." (which also results in converse holding since this is a definition)
}

{
1450. (If a prophet is) . . . born of women . . . (then that prophet) . . . is not a greater prophet than John the Baptist . . . (Disc) (GA/MR-GR) (INFERRED) (Deductive) - NOT in Matthew, Mark, or John
Contrapositive: verseF? – From statement attributed to Christ as, ". . . This is the heir . . ."

(If a prophet is) . . . not a greater prophet than John the Baptist . . . (then that human is) . . . born of women . . . (Disc) (GA-MA/GR) (INFERRED) (Deductive) - NOT in Matthew, Mark, or John
Contrapositive: verseF? – From statement attributed to Christ as, ". . . This is the heir . . ." (which also results in converse holding since this is a definition)
}
{
1451. (If a thing is) . . . the leaven of the Pharisees . . . (, then that thing) . . . is hypocrisy. (Disc) (MAMR) (INFERRED) (Neither inductive nor deductive) - NOT in Matthew, Mark, or John
Contrapositive: verseF? – From statement attributed to Christ as, ". . . the leaven of the Pharisees, which is hypocrisy."

(If a thing) . . . is hypocrisy . . . (, then that thing is) . . . the leaven of the Pharisees. (Disc) (GAGR) (INFERRED) (Neither inductive nor deductive) - NOT in Matthew, Mark, or John
Contrapositive: verseF? – From statement attributed to Christ as, "the leaven of the Pharisees, which is hypocrisy . . ." (which also results in converse holding since this is a definition)
}

1452. (If a thing is) . . . the life (of a human) . . . (, then that thing) . . . is more than meat. (Disc) (MAMR) (INFERRED) (Deductive) - NOT in Matthew, Mark, or John
Contrapositive: verseF? – From statement attributed to Christ as, "The life is more than meat . . ." {Note: The converse is assumed to be FALSE since there is no statement to address this in the words attributed to Christ, and acoording to the formal rules of human reasoning, anything more than meat (which

in the contextm "meat" is food for humans) is not necessarily the life (of a human).}

{
1453. (If a thing) . . . is highly esteemed among men . . . (,then that thing) . . . is abomination in the sight of God. (Disc) (GAGR) (INFERRED) (Deductive) Contrapositive: verseF? - NOT in Matthew, Mark, or John - From statement attributed to Christ, ". . . that which is highly esteemed among men is abomination in the sight of God."

(If a thing) . . . is abomination in the sight of God . . . (,then that thing) . . . is highly esteemed among men . . . (Disc) (GAGR) (INFERRED) (Deductive) Contrapositive: verseF? - From statement attributed to Christ, ". . . that which is abomination in the sight of God is highly esteemed among men. . . ." (which also results in converse holding since this is a definition)
}
{
1454. (If a thing is) . . . my (God's) house . . . (,then that thing) . . . is the house of prayer (Disc) (GAGR) (INFERRED) (Neither inductive nor deductive) – NOT in Matthew, Mark, or John
Contrapositive: verseF? - From statement attributed to Christ, ". . . My house is the house of prayer . . ."

(If a thing) . . . is the house of prayer . . . (,then that thing is) . . . my (God's) house . . . (Disc) (GAGR) (INFERRED) (Neither inductive nor deductive) – NOT in Matthew, Mark, or John
Contrapositive: verseF? - From statement attributed to Christ, "My house is the house of prayer . . ." (which also results in converse holding since this is a definition SINCE THE PHRASE, "THE HOUSE OF PRAYER" IS USED)
}
{
1455. (If a thing) . . . is born of the flesh . . . (,then that thing) . . . is flesh (Disc) (GAGR) (INFERRED) (Deductive) – NOT in Matthew, Mark, or Luke
Contrapositive: verseF? - From statement attributed to Christ, ". . . that which is born of the flesh is flesh . . ."

(If a thing) . . . is flesh . . . (,then that thing is) . . . is born of the flesh . . . (Disc) (GAGR) (INFERRED) (Deductive) – NOT in Matthew, Mark, or John
Contrapositive: verseF? - From statement attributed to Christ, ". . . that which is born of the flesh is flesh . . ." (which also results in converse holding since this is a definition)
}

1456. (If a thing) . . . is born of the spirit . . . (,then that thing) . . . is spirit (Disc) (GAGR) (INFERRED) (Deductive) – NOT in Matthew, Mark, or Luke Contrapositive: verseF? - From statement attributed to Christ, ". . . that which is born of the spirit is spirit . . ." {KEY NOTE: The converse of this statement is FALSE using the words attributed to Christ since Christ stated, "God the Father is a spirit . . .", according to # 1458 (John 4:24), but cannot be said to have been "born" as # 1459 (John 17: 3) assumes that God (the Father) is eternal in order for eternal life to be directly connected to knowing God (the Father).} KEY EXCEPTION TO THE STRICT RULES OF FORMAL REASONING

1457. (If a thing is) . . . God (the Father) . . . (,then that thing) . . . is (part of the group of all things that have the trait of being) . . . a spirit . . . (Samaritan woman) (GAGR) (INFERRED) (Deductive)
Contrapositive: verseF? - NOT in Matthew, Mark, or Luke - From statement attributed to Christ, "God is a spirit . . ." (KEY NOTE: The converse of this statement is FALSE using the words attributed to Christ since Christ stated several times in John 14–17 that the Holy Spirit is a spirit, but is not the same as God (the Father) or Christ and since Christ indicates that man has a 'spirit in Matthew 22: 36, but also implies in many passages, including John 3: 16 that man is different from God (the Father).

{
1458. (If a thing is) . . . life eternal . . . (,then that thing is such) . . . that they might know thee the only true God, and Jesus Christ, whom thou hast sent." (Disc) (GA/GR-MR) (INFERRED) (Deductive)
Contrapositive: verseF? – NOT in Matthew, Mark, or Luke From statement attributed to Christ, "This is life eternal, that they might know thee the only true God, and Jesus Christ, whom thou hast sent." (

(If a thing is such) . . . that they might know thee the only true God, and Jesus Christ, whom thou hast sent . . . (,then that thing is) . . . .eternal life . . ." (Disc) (GA-MA/GR) (INFERRED) (Deductive)
Contrapositive: verseF? - From statement attributed to Christ, "This is life eternal, that they might know thee the only true God, and Jesus Christ, whom thou hast sent." (which also results in converse holding since this is a definition)
}
{
1459. (If a thing is) . . . thee (God the Father) . . . (,then that thing is) . . . the only true God." (Disc) (GAGR) (INFERRED) (Neither inductive nor deductive)

Contrapositive: verseF? - NOT in Matthew, Mark, or Luke - From statement attributed to Christ, "... thee the only true God ..."

(If a thing is) ... the only true God ... (,then that thing is) ... thee (God the Father)" (Disc) (GAGR) (INFERRED) (Neither inductive nor deductive)
Contrapositive: verseF? - From statement attributed to Christ, "... thee the only true God ..." (which also holds for the converse since this is a defin-tion)
}
{
1460. (If a thing is) ... this (that light is come into the world, and men loved darkness rather than light, because their deeds were evil) ... (,then that thing is) ... the condemnation (of God the Father)." (Nicodemus) (GA-MA/GR) (INFERRED) (Deductive)
Contrapositive: verseF? – NOT in Matthew, Mark, or Luke From statement attributed to Christ, "And this is the condemnation, that light is come into the world, and men loved darkness rather than light, because their deeds were evil."

(If a thing is) ... the condemnation (of God the Father) ... (,then that thing is such).. this (that light is come into the world, and men loved darkness rather than light, because their deeds were evil) ..." (Nicodemus) (GA/GR-MR) (INFERRED) (Deductive)
Contrapositive: verseF? - From statement attributed to Christ, "And this is the condemnation, that light is come into the world, and men loved darkness rather than light, because their deeds were evil." (which also holds for the converse since this is a defintion)
}
{
1461. ... because their deeds were evil, (then) ... men loved darkness rather than light ..." (Disc) (MAMR) (INFERRED) (Deductive)
Contrapositive: verseF? – NOT in Matthew, Mark, or Luke - The converse also holds as an inferred, conditional statement
}

1462. (If a man is) ... whom thou now hast (that is, the man with whom the Samaritan woman at the well was living at the time she spoke to Christ) ... (then that man) ... is not thy (the Samaritan woman at the well) husband ... (Samaritan woman at the well) (MAMR) (INFERRED) (Neither inductive nor deductive)

Contrapositive: verse? – NOT in Matthew, Mark, or Luke—The converse does not hold here.

1463. (If a thing is such that its) . . . hour cometh . . . (and is occurring) . . . now (at the time Christ spoke to the Samaritan woman at the well) . . . (,then that thing is such that) . . . the true worshippers shall worship the Father in spirit and in truth." (Samaritan woman at the well) (GAMR) (INFERRED) (Neither inductive nor deductive)
Contrapositive: verse? – NOT in Matthew, Mark, or Luke - The converse cannot be addressed by the statements attributed to Christ in the Holy Bible, and since a future conditional cannot be assessed relative to a present conclusion (as required since 'now' is also used here), the rules of formal human reasoning do not apply in such cases.

{
1464. (If a thing is) . . . my (Christ's) meat . . . (, then that thing) is to do the will of him (God the Father) that sent me (Christ) . . . (Disc) (GAGR) (INFERRED) (Neither inductive nor deductive)
Contrapositive: verse? - NOT in Matthew, Mark, or Luke

(If a thing is) . . . my (Christ's) meat . . . (, then that thing) is to do the will of him (God the Father) that sent me (Christ) . . . (Disc) (GAGR) (INFERRED) (Neither inductive nor deductive)
Contrapositive: verse? – The converse holds since this is a definition.
}

1465. (If a thing is such that) . . . the hour is coming . . . (and is occurring) . . . now is (at the time Christ spoke to the Jews after healing the lame man at the pool of Bethsada) . . . (,then that thing is such that) . . . the dead shall hear the voice of the Son of God: and they that hear shall live." (Relig) (GA/GR-MR) (INFERRED) (Neither inductive nor deductive)
Contrapositive: verse? – NOT in Matthew, Mark, or Luke- The converse cannot be addressed by the statements attributed to Christ in the Holy Bible, and since a future conditional cannot be assessed relative to a present conclusion (as required since 'now' is also used here), the rules of formal human reasoning do not apply in such cases.

1466. (if a human is) . . . Moses in whom ye (the Jewish religious leaders) 'faith' . . . (,then that human) .. is one that accuseth you (the Jewish religious leaders) . . . (Relig) (MAMR) (INFERRED) (Neither inductive nor deductive)

Contrapositive: verseF? – NOT in Matthew, Mark, or Luke - The converse does not hold here.

{
1467. (If a thing is) . . . this (that ye believe on him whom he hath sent) . . . (, then that thing) . . . is the work of God (for ye as non-believers) . . . (Mult) (GA-MA/GR) (INFERRED) (Deductive)
Contrapositive: verse? - NOT in Matthew, Mark, or Luke

(If a thing) . . . is the work of God (for 'ye' as the general population, which includes non-believers) . . . (, then that thing is) . . . this (that ye believe on him whom he hath sent) . . . (Mult) (GA/MR-GR) (INFERRED) (Deductive)
Contrapositive: verseF? – The converse holds if the unstated condition is used from the given context to the "work of God" (to essentially unbelievers) since the context is clear that those to whom Christ is speaking in these statements are those who do not believe and do not include the disciples as per John 6: 24.
}

1468. (If a thing is) . . . the bread of God . . . (, then that thing) . . . is he which cometh down from heaven, and giveth life unto the world. (Mult) (GAGR) (INFERRED) (Deductive)
Contrapositive: verse? – NOT in Matthew, Mark, or Luke—see # 1602 and # 1400 for similar instances; The converse does not hold since 'bread from God' could include that which Christ referenced as the 'manna' sent by God to the children of Israel in the 'wilderness' from the Old Testament

{
1469. (If a thing is) . . . the bread that I (Christ) will give . . . , then that thing is) . . . my (Christ's) flesh. . . . (Mult) (GAGR) (INFERRED) (Neither inductive nor deductive)
Contrapositive: verseF? – NOT in Matthew, Mark, Luke – From statement attributed to Christ, ". . . the bread that I will give is my flesh . . ."

(If a thing is) . . . my (Christ's) flesh . . . , (then that thing is) . . . the bread that I (Christ) will give. (Mult) (GAGR) (INFERRED) (Neither inductive nor deductive)
Contrapositive: verseF? – NOT in Matthew, Mark, Luke – From statement attributed to Christ, ". . . the bread that I will give is my flesh . . ." (which also results in converse holding since this is a definition {Note: This statement is not applied to the discussion regarding ". . . this (the bread used when Christ

instituted the Lord's Supper) is my body . . ." since the time in the future at which Christ "will give" this "bread" is not explicitly identified (i.e., whether the future time is at the institution of the Lord's Supper, at the time of His death, or otherwise), which is important since it is generally agreed by Bible scholars that the Old Testament Scriptures only explicitly presented the bread from the celebration of the Passover meal to be specific to "feast of the unleavened bread" eatern during the quick departure of the Jewish people from Egypt in chapter 12 of the book of Exodus.}
}
{
1470. (If a thing is) . . . my flesh . . . , then that thing) . . . is meat indeed. . . .
(Mult) (GAGR) (INFERRED) (Neither inductive nor deductive)
Contrapositive: verseF? – NOT in Matthew, Mark, Luke – From statement attributed to Christ, ". . . my flesh is meat indeed . . ."

(If a thing) . . . is meat indeed . . . , (then that thing is) . . . my flesh . . . . (Mult) (GAGR) (INFERRED) (Neither inductive nor deductive)
Contrapositive: verseF? – NOT in Matthew, Mark, Luke – From statement attributed to Christ, ". . . my flesh is meat indeed . . ." (which also results in converse holding since this is a definition {Note: See # 1469}
}
{
1471. (If a thing is) . . . my blood . . . , then that thing) . . . is drink indeed. . . .
(Mult) (GAGR) (INFERRED) (Neither inductive nor deductive)
Contrapositive: verseF? – NOT in Matthew, Mark, Luke – From statement attributed to Christ, ". . . my blood is drink indeed . . ."

(If a thing) . . . is drink indeed . . . , (then that thing is) . . . my blood . . . .
(Mult) (GAGR) (INFERRED) (Neither inductive nor deductive)
Contrapositive: verseF? – NOT in Matthew, Mark, Luke – From statement attributed to Christ, ". . . my blood is drink indeed . . ." (which also results in converse holding since this is a definition {Note: This statement is not applied to the discussion regarding ". . . this (the bread used when Christ instituted the Lord's Supper) is my blood of the new testament . . ." since the time in the future Christ's "blood" is not explicitly identified (i.e., whether the future time is at the institution of the Lord's Supper, at the time of His death, or otherwise), which is important since it is generally agreed by Bible scholars that the Old Testament Scriptures only explicitly presented the blood from the celebration of the Passover to be specific to the blood of the lamb (young sheep) used on the upper and side posts of the houses in which the Passover meal was to be eaten as per chapter 12 of the book of Exodus, though it is true that

others prior to this time, such as John the Baptist, are indicated in the Holy Bible (for example, see John 1: 29) when he referred to Christ as "the Lamb of God, which taketh away the sin of the world" .}
}

{
1472. (If a thing) . . . is the spirit . . . , then that thing is) . . . that (which) quickeneth. . . . (Mult) (GAGR) (INFERRED) (Deductive)
Contrapositive: verseF? – NOT in Matthew, Mark, Luke – From statement attributed to Christ, ". . . It is the spirit that quickeneth . . ."

(If a thing is) . . . that (which) quickeneth . . . (, then that thing) . . . is the spirit . . . . (Mult) (GAGR) (INFERRED) (Deductive)
Contrapositive: verseF? – NOT in Matthew, Mark, Luke – From statement attributed to Christ, ". . . It is the spirit that quickeneth . . ." (which also results in converse holding since this is a definition)
}
{
1473. (If things are) . . . the words that I (Christ) speak unto you . . . (, then those things) . . . are spirit. . . . (Mult) (GAGR) (INFERRED) (Neither inductive nor deductive)
Contrapositive: verseF? – NOT in Matthew, Mark, Luke – From statement attributed to Christ, ". . . the words that I speak unto you, they are spirit . . ."

(If those things) . . . are spirit . . . (, then those things are) . . . the words that I (Christ) speak unto you . . . . (Mult) (GAGR) (INFERRED) (Neither inductive nor deductive)
Contrapositive: verseF? – NOT in Matthew, Mark, Luke – From statement attributed to Christ, ". . . the words that I (Christ) speak unto you, they are spirit . . ." (which also results in converse holding since this is a definition)
}
{
1474. (If things are) . . . the words that I (Christ) speak unto you . . . (, then those things) . . . are life. . . . (Mult) (GAGR) (INFERRED) (Neither inductive nor deductive)
Contrapositive: verseF? – NOT in Matthew, Mark, Luke – From statement attributed to Christ, ". . . the words that I speak unto you, they are life . . ."

(If those things) . . . are life . . . (, then those things are) . . . the words that I (Christ) speak unto you . . . . (Mult) (GAGR) (INFERRED) (Neither inductive nor deductive)

Contrapositive: verseF? – NOT in Matthew, Mark, Luke – From statement attributed to Christ, ". . . the words that I (Christ) speak unto you, they are life . . ." (which also results in converse holding since this is a definition)
}
{
1475. (If a thing is) . . . my (Christ's) doctrine . . . (, then that thing) . . . is not mine (Christ's). . . . (Mult) (GAGR) (INFERRED) (Neither inductive nor deductive)
Contrapositive: verseF? – NOT in Matthew, Mark, Luke – From statement attributed to Christ, ". . . My doctrine is not mine . . ."

(If a thing) . . . is not mine (Christ's) . . . (, then that thing is) . . . my (Christ's) doctrine . . . . (Mult) (GAGR) (INFERRED) (Neither inductive nor deductive)
Contrapositive: verseF? – NOT in Matthew, Mark, Luke – From statement attributed to Christ, ". . . My doctrine is not mine . . ." (which also results in converse holding since this is a definition) – THIS IS AN EXCEPTION
}
{
1476. (If a thing is) . . . my (Christ's) doctrine . . . (, then that thing is) . . . . his (doctrine) that sent me . . . (Mult) (GAGR) (INFERRED) (Neither inductive nor deductive)
Contrapositive: verseF? – NOT in Matthew, Mark, Luke – From statement attributed to Christ, ". . . My doctrine is not mine, but his that sent me . . ."

(If a thing) . . . is not mine (Christ's) . . . (, then that thing is) . . . my (Christ's) doctrine . . . . (Mult) (GAGR) (INFERRED) (Neither inductive nor deductive)
Contrapositive: verseF? – NOT in Matthew, Mark, Luke – From statement attributed to Christ, ". . . My doctrine is not mine, but his that sent me . . ." (which also results in converse holding since this is a definition)
}
{
1477. (If a thing is) . . . truth . . . (, then that thing is) . . . .not . . . in him (the devil) . . . (Mult) (GAGR) (INFERRED) (Neither inductive nor deductive)
Contrapositive: verseF? – NOT in Matthew, Mark, Luke – From statement attributed to Christ, ". . . there is no truth in him (the devil) . . ."

(If a thing is) . . . not . . . in him (the devil) . . . (, then that thing is) . . . truth . . . . (Mult) (GA__) (INFERRED) (Neither inductive nor deductive)

Contrapositive: verseF? – NOT in Matthew, Mark, Luke – From statement attributed to Christ, "... there is no truth in him (the devil) ..." (which also results in converse holding since this is a definition)
}

1478. (If a thing is) ... he (the devil) ... (, then that thing is) ... a liar ... (Mult) (GA__) (INFERRED) (Neither inductive nor deductive)
Contrapositive: verseF? – NOT in Matthew, Mark, Luke – From statement attributed to Christ, "... he (the devil) is a liar ..." – The converse does not hold since there are liars who are not the devil, but this cannot be proven with the statements attributed to Christ

{
1479. (If a thing is) ... he (the devil) ... (, then that thing is) ... the father of it (liars) ... (Mult) (GA__) (INFERRED) (Neither inductive nor deductive)
Contrapositive: verseF? – NOT in Matthew, Mark, Luke – From statement attributed to Christ, "... he (the devil) is the father of it (liars) ..."

(If a thing is) ... the father of it (liars) ... (, then that thing is) ... he (the devil) .... (Mult) (GAGR) (INFERRED) (Neither inductive nor deductive)
Contrapositive: verseF? – NOT in Matthew, Mark, Luke – From statement attributed to Christ, "... he (the devil) is the father of it (liars) ..." (which also results in converse holding since this is a definition)
}

1480. (if one) ... honoureth not the Son (Christ) (,then that one) ... honoureth not the Father which sent him. (Relig) (MAMR) (INFERRED) (Neither inductive nor deductive)
Contrapositive: verse? – NOT in Matthew, Mark, or Luke – This is an example of the contrapositive being explicitly stated as true in words attributed to Christ. Specifically, # 1480 is the contrapositive of # 1055.

1481. (If a thing is) ... my (Christ's) Father ... (, then that thing is such) ... that honoureth me ... (Mult) (GAGR) (INFERRED) (Neither inductive nor deductive)
Contrapositive: verse? – NOT in Matthew, Mark, Luke – From statement attributed to Christ, "... it is my Father that honoureth me ..." – The converse does not hold since "honour" (as oppsed to "nonoureth not" in # 1480) of Christ is not explicitly addressed as other than a conclusion in the statements attributed to Christ

1482. (If a thing is) . . . the Son of God . . . (, then that thing is such) . . . he (Jesus) that talketh to thee (the blind man who was healed by Jesus) . . . (Indiv-blind man) (GAGR) (INFERRED) (Neither inductive nor deductive)
Contrapositive: verse? – NOT in Matthew, Mark, Luke – From statement attributed to Christ, ". . . it (the Son of God) is he (Jesus) that talketh to thee . . ."
{
1483. (If things are) . . . other sheep I (Christ) have . . . (, then those things are) . . . not of this fold . . . (Mult) (GAGR) (INFERRED) (Neither inductive nor deductive)
Contrapositive: verse? – NOT in Matthew, Mark, Luke – From statement attributed to Christ, ". . . other sheep I have which are not of this fold . . ."

(If things are) . . . not of this fold . . . (, then those things are) . . . other sheep I (Christ) have . . . (Mult) (GAGR) (INFERRED) (Neither inductive nor deductive)
Contrapositive: verse? – NOT in Matthew, Mark, Luke – From statement attributed to Christ, ". . . other sheep I have which are not of this fold . . ." (which holds as the converse since this is a definition)
}
{
1484. (If a thing is) . . . My Father, which gave them (sheep) me . . . (, then that thing) . . . is greater than all . . . (Mult) (GAGR) (INFERRED) (Neither inductive nor deductive)
Contrapositive: verse? – NOT in Matthew, Mark, Luke – From statement attributed to Christ, ". . . My Father, which gave them me, is greater than all . . ."

(If a thing) . . . is greater than all . . . (, then that thing is) . . . My Father, which gave them me . . . (Mult) (GAGR) (INFERRED) (Neither inductive nor deductive)
Contrapositive: verse? – NOT in Matthew, Mark, Luke – From statement attributed to Christ, ". . . My Father, which gave them me, is greater than all . . ." (which holds as the converse since this is a definition)
}

1485. (If a thing is) . . . This sickness (for Lazarus) . . . (, then that thing) . . . is not unto death . . . (Mult) (GA/MR-GR) (INFERRED) (Neither inductive nor deductive)
Contrapositive: verse? – NOT in Matthew, Mark, or Luke

1486. (If a human is) . . . Lazarus) . . . (, then that thing at least at the time Christ spoke to his disciples before going to see Lazarus' family) . . . is dead . . . (Mult) (GA/MR-GR) (INFERRED) (Neither inductive nor deductive)

Contrapositive: verse? – NOT in Matthew, Mark, or Luke – This is a violation of # 1485 given the use of "unto", though such can be seen as "not to end in death" in all other major translations

1487. (if a thing) . . . is his (God the Father's) commandment . . . (, then that thing) . . . is life everlasting . . . (Disc) (GAGR) (INFERRED) (Neither inductive nor deductive)
Contrapositive: verse? – NOT in Matthew, Mark, or Luke – From a statement attributed to Christ, ". . . his commandment is life everlasting . . ."

(if a thing) . . . is life everlasting (, then that thing is) . . . his (God the Father's) commandment . . . (Disc) (GAGR) (INFERRED) (Neither inductive nor deductive)
Contrapositive: verse? – NOT in Matthew, Mark, or Luke – From a statement attributed to Christ, ". . . his commandment is life everlasting . . ."

{
1488. (If a human is) . . . the one to whom I shall give a sop (when I have dipped it at the celebration of the Passover meal on the night of Christ's Trials before His Crucifixion) (,then that human) . . . is . . . he (whom shall betray me). (John the disciple) (GAMR) (INFERRED) (Neither inductive nor deductive)
Contrapositive: verse? – NOT in Matthew, Mark, or Luke

(If a human is) . . . he (whom shall betray me) . . . (, then that human is) . . . the one to whom I shall give a sop (when I have dipped it at the celebration of the Passover meal on the night of Christ's Trials before His Crucifixion). (John the disciple) (MAGR) (INFERRED) (Neither inductive nor deductive)
Contrapositive: verse? – NOT in Matthew, Mark, or Luke – The converse holds given that this is a definition
}
{
1489. (If a thing is) . . . the word which ye (the disciples) hear me speak . . . (, then that word) . . . is not mine, but the Father's which sent me. . . . (Disc) (GAGR) (INFERRED) (Neither inductive nor deductive)
Contrapositive: verseF? – NOT in Matthew, Mark, Luke – From statement attributed to Christ, ". . . . . ."

(If a thing is a word that) . . . is not mine, but the Father's which sent me . . . (, then that thing is) . . . the word which ye (the disciples) hear me speak . . . . (Disc) (GAGR) (INFERRED) (Neither inductive nor deductive)

Contrapositive: verseF? – NOT in Matthew, Mark, Luke – From statement attributed to Christ, "... My doctrine is not mine ..." (which also results in converse holding since this is a definition) – THIS IS AN EXCEPTION
}
{
1490. (If a thing is God) my Father ... (, then that thing) ... is the husbandman. ... (Disc) (GAGR) (INFERRED) (Neither inductive nor deductive)
Contrapositive: verse? – NOT in Matthew, Mark, Luke – From statement attributed to Christ, "... my Father is the husbandman ..."

(If a thing) ... the husbandman ..., (then that thing is God) my Father .... (Disct) (GAGR) (INFERRED) (Neither inductive nor deductive)
Contrapositive: verse? – NOT in Matthew, Mark, Luke – From statement attributed to Christ, "... my Father is the husbandman ..." (which also results in converse holding since this is a definition {Note: See # 1469}
}
{
1491. (If a thing is) my (Christ's) commandment ... (, then that thing is) ... That ye love one another, as I have loved you. (Disc) (GAMR) (INFERRED) (Neither inductive nor deductive)
Contrapositive: verse? – NOT in Matthew, Mark, Luke – From statement attributed to Christ, "... This is my commandment, That you love one another, as I have loved you ..."

(If a thing is) ... That ye love one another, as I have loved you ..., (then that thing is) my (Christ's) commandment .... (Disc) (MAGR) (INFERRED) (Neither inductive nor deductive)
Contrapositive: verse? – NOT in Matthew, Mark, Luke – From statement attributed to Christ, "... This is my commandment, That you love one another, as I have loved you ..." (which also results in converse holding since this is a definition {Note: See # 1469}
}
{
1492. (If a thing is) thy (God the Father's) word ... (, then that thing) ... truth. ... (Disc) (GAGR) (INFERRED) (Neither inductive nor deductive)
Contrapositive: verseF? – NOT in Matthew, Mark, Luke – From statement attributed to Christ, "... thy word is truth ..."

(If a thing) ... truth ..., (then that thing is) thy (God the Father's) word .... (Disct) (GAGR) (INFERRED) (Neither inductive nor deductive)

Contrapositive: verseF? – NOT in Matthew, Mark, Luke – From statement attributed to Christ, "... they word is truth ..." (which also results in converse holding since this is a definition {Note: See # 1469}
}
{
1493. (If a human is) ye (any disciple of Christ) ... (,then that human) ... is the salt of the earth. (disciples) (MA-GA/MR-GR) (INFERRED) (Deductive)
Contrapositive: verse? – NOT in Matthew, Mark, or Luke – From statement attributed to Christ, "Ye are the salt of the earth ..."

(If a human) ... is the salt of the earth ... (, then that human is) ... ye (any disciple of Christ). (disciples) (MA-GA/MR-GR) (INFERRED) (Deductive)
Contrapositive: verse? – NOT in Matthew, Mark, or Luke – – From statement attributed to Christ, "Ye are the salt of the earth ..."—The converse holds given that this is a definition.
}
{
1494. (If a human is) ye (any disciple of Christ) ... (,then that human) ... is the light of the world. (disciples) (MA-GA/MR-GR) (INFERRED) (Deductive)
Contrapositive: verse? – NOT in Mark, Luke, or John – From statement attributed to Christ, "Ye are the salt of the earth ..."

The converse holds given # 1402, #1101, and # 1103 in which Christ is attributed as saying, "I am the light of the world."
}

1495. PART OF A DUAL INCLUSION: "... (a disciple) dwelleth in me ..." so that (if one is a disciple, then that one) ... dwelleth in me ... – John 6: 56

1496. PART OF A DUAL INCLUSION: "I (Christ dwelleth) in him (a disciple)." So that (if one is) ... I (Christ) ... (, then that one dwelleth) in him (a disciple). - John 6: 56
{
1497. (If a set of humans are) ye (hypocrites) ... (,then those humans) ... are full of hypocrisy. (Relig) (MAMR) (INFERRED) (Deductive)

Contrapositive: verse? – NOT in Mark, Luke, or John – From statement attributed to Christ, "Ye are ... full of hypocrisy ..."

(If a set of humans) . . . are full of hypocrisy . . . (,then those humans are) . . . ye (hypocrites). (Relig) (MAMR) (INFERRED) (Deductive)
Contrapositive: verse? – NOT in Mark, Luke, or John – From statement attributed to Christ, "Ye are . . . full of hypocrisy . . ." – The converse holds since this is a definition.
}
1498. (If a set of humans) . . . are the children of them which killed the prophets (,then those humans are) . . . ye (hypocrites). (Relig) (MAMR) (INFERRED) (Deductive)
Contrapositive: verse? – NOT in Mark, Luke, or John – From statement attributed to Christ, "Ye are . . . full of hypocrisy . . ." – The converse does not hold since there are children of them which killed the prophets which are not hypocrites, scribes, or Pharisees.
{
1499. (If a set of humans are) . . . ye (disciples of Christ) . . . (,then those humans) . . . belong to Christ. (Relig) (MA/MR-GR) (INFERRED) (Deductive)
Contrapositive: verse? – NOT in Mark, Luke, or John – From statement attributed to Christ, "Ye belong to Christ . . ."

(If a set of humans) . . . belong(s) to Christ . . . (,then those humans are) . . . ye (disciples of Christ). (Relig) (MAMR) (INFERRED) (Deductive)
Contrapositive: verse? – NOT in Matthew, Luke, or John – From statement attributed to Christ, "Ye belong to Christ . . ." – The converse holds since this is a definition.
}
1500. (If a set of humans are) . . . ye (Pharisees) . . . (,then those humans) . . . are they which justify yourselves before men. (Relig) (MA/MR-GR) (INFERRED) (Neither inductive nor deductive)
Contrapositive: verse? – NOT in Matthew, Mark, or John – From statement attributed to Christ, "Ye are they which justify yourselves before men . . ."— The converse does not holds since one can justify oneself before men without being a Pharisee.
{
1501. (If a set of humans are) ye (one of the 11 original apostles of Christ) . . . (,then those humans) . . . are they which have continued with me in my temptation. (Disciples) (MA-GA/MR) (INFERRED) (Neither inductive nor deductive)
Contrapositive: verse? – NOT in Matthew, Mark, or John – From statement attributed to Christ, "Ye are they which have continued with me in my temptation . . ."

(If a set of humans) . . . are they which have continued with me in my temptation . . . (,then those humans are) . . . ye (one of the 11 original apostles of Christ). (Disciples) (MA/GR-/MR) (INFERRED) (Neither inductive nor deductive)
Contrapositive: verse? – NOT in Matthew, Mark, or John – From statement attributed to Christ, "Ye are they which have continued with me in my temptation . . ." - The converse holds since this is a definition."
}
1502. (If a set of humans are) ye (one of the 11 original apostles of Christ and two (2) disciples of Christ, including Cleopas) . . . (,then those humans) . . . are (one of the) witnesses of these things {"(Christ's) suffer(ing) and ris(ing) from the dead the third day"}. (Disciples) (MA-GA/MR) (INFERRED) (Neither inductive nor deductive)
Contrapositive: verse? – NOT in Matthew, Mark, or John – From statement attributed to Christ, "Ye are witnesses of these things . . ."

The converse does not hold since there were other witnesses, including Mary Magdalene, which are not in this group."

1503. (If a set of humans are) . . . ye (Pharisees) . . . (,then those humans) . . . are from beneath. (Relig) (MAMR) (INFERRED) (Neither inductive nor deductive)
Contrapositive: verse? – NOT in Matthew, Mark, or Luke – From statement attributed to Christ, "Ye are they which justify yourselves before men . . ."—The converse does not holds since one can be "from beneath", such as unbelieving Gentiles, without being a Pharisee.

1504. (If a set of humans are) . . . ye (Pharisees) . . . (,then those humans) . . . are of this world. (Relig) (MAMR) (INFERRED) (Neither inductive nor deductive)
Contrapositive: verse? – NOT in Matthew, Mark, or Luke – From statement attributed to Christ, "Ye are of this world . . ."—The converse does not holds since one can be "of this world" without being a Pharisee.

1505. (If a set of humans are) . . . ye (Jews who at first believed on Christ, but who then wanted to kill Christ) . . . (,then those humans) . . . are Abraham's seed (the biological descendants of Abraham). (Relig) (MAMR) (INFERRED) (Deductive)
Contrapositive: verse? – NOT in Matthew, Mark, or Luke – From statement attributed to Christ, "Ye are Abraham's seed . . ."

The converse does not hold since # 1137 states directly that there were other of "Abraham's children" who did not seek to kill Christ.

1506. (If a set of humans are) . . . ye (Jews who at first believed on Christ, but who then wanted to kill Christ) . . . (,then those humans) . . . are of your father the devil. (Relig) (MAMR) (INFERRED) (Deductive)
Contrapositive: verse? – NOT in Matthew, Mark, or Luke – From statement attributed to Christ, "Ye are of your father the devil . . ."

The converse does not hold since there are others whose "father is the devil", but who are Gentiles.

1507. (If a set of humans are) . . . ye (Jews who at first believed on Christ, but who then wanted to kill Christ) . . . (,then those humans) . . . are not of God. (Relig) (MAMR) (INFERRED) (Deductive)
Contrapositive: verse? – NOT in Matthew, Mark, or Luke – From statement attributed to Christ, "Ye are not of God . . ."

The converse does not hold since there are others whose "father is the devil", but who are Gentiles.

1508. (If a set of humans are) . . . ye (Jews who do not believe on Christ) . . . (,then those humans) . . . are not of my sheep. (Relig) (MAMR) (INFERRED) (Deductive)
Contrapositive: verse? – NOT in Matthew, Mark, or Luke – From statement attributed to Christ, "Ye are not of my sheep . . ."

The converse does not hold since there were Gentiles who were not of Christ's sheep.

1509. (If a set of humans are) ye (one of the 11 original apostles of Christ) . . . (,then those humans) . . . are clean . . . . (Disciples) (MA-GA/MR) (INFERRED) (Neither inductive nor deductive)
Contrapositive: verse? – NOT in Matthew, Mark, or Luke – From statement attributed to Christ, "Ye are clean, but not all . . ."

The converse does not hold since there are others who are "clean" (in Christ's view), but who were not one of the 11 original Apostles.

1510. (If a set of humans are) ye (one of the 11 original apostles of Christ) . . . (,then those humans) . . . are my friends . . . . (Disciples) (MA-GA/MR) (INFERRED) (Neither inductive nor deductive)
Contrapositive: verse? – NOT in Matthew, Mark, or Luke – From statement attributed to Christ, "Ye are my friends . . ."

The converse does not hold since there are others who are "friends of Christ'" (in Christ's view), but who were not one of the original Apostles.

1511. (If a set of humans are) ye (one of the 11 original apostles of Christ) . . . (,then those humans) . . . are not of this world . . . . (Disciples) (MA-GA/MR) (INFERRED) (Neither inductive nor deductive)
Contrapositive: verse? – NOT in Matthew, Mark, or Luke – From statement attributed to Christ, "Ye are not of this world . . ."

The converse does not hold since there are others who are "not of this world" (in Christ's view), but who were not one of the original Apostles.

1512. (If a set of humans are) ye (one of the 11 original apostles of Christ) . . . (,then those humans) . . . are clean . . . . (Disciples) (MA-GA/MR) (INFERRED) (Neither inductive nor deductive)
Contrapositive: verse? – NOT in Matthew, Mark, or John – From statement attributed to Christ, "Ye are clean, but not all . . ."

The converse does not hold since there are others who are "clean" (in Christ's view), but who were not one of the original Apostles.

{
1513. (If a set of humans are) ye (one of the 11 original apostles of Christ) . . . (,then those humans) . . . have been (are) with me from the beginning (of my earthly ministry) . . . . (Disciples) (MA-GA/MR) (INFERRED) (Neither inductive nor deductive)
Contrapositive: verse? – NOT in Matthew, Mark, or John – From statement attributed to Christ, ". . . ye have been with me from the beginning . . ."

(If a set of humans) . . . have been (are) with me from the beginning (of my earthly ministry) . . . (,then those humans) . . . ye (one of the 11 original apostles of Christ. (Disciples) (MA-GA/MR) (INFERRED) (Neither inductive nor deductive)
Contrapositive: verse? – NOT in Matthew, Mark, or John – From statement attributed to Christ, ". . . ye have been with me from the beginning . . ."

The converse holds since this is a definition.
}

1514. (If a human is) . . . I (Jesus Christ) (,then that human) . . . am (is) not of this world . . . . (Relig) (MA-GA/MR) (INFERRED) (Neither inductive nor deductive)

Contrapositive: verse? – NOT in Matthew, Mark, or Luke – From statement attributed to Christ, "I am not of this world . . ."

The converse does not hold since there are others who are "not of this world" (in Christ's view), but are not Jesus Christ.

1515. (If it is) . . . now (at the time of Christ's baptism by John) . . . (then) . . . Suffer it to be so (Christ coming to John for baptism) . . . (John the Baptist) (GAMR) (INFERRED) (Neither inductive nor deductive)

Contrapositive: verseF? – NOT in Mark, Luke, or John

The converse does not hold.

1516. (since) . . . thus it becometh us to fulfill all righteousness . . . (and) (If it is) . . . now (at the time of Christ's baptism by John) . . . (then) . . . Suffer it to be so (Christ coming to John for baptism) . . . (John the Baptist) (GAMR) (INFERRED) (Neither inductive nor deductive)

Contrapositive: verseF? – NOT in Mark, Luke, or John

The converse does not hold.

1517. . . . (since) the kingdom of heaven is at hand . . . (then) . . . Repent . . . (Mult) (GAMR) (INFERRED) (Neither inductive nor deductive)

Contrapositive: verseF? – NOT in Mark, Luke, or John

The converse does not hold.

1518. (if) you are persecute(d) and (have men to) say all manner of evil against you, for my sake, (then) Rejoice, and be exceeding glad: (since) . . . great is your reward in heaven . . . (MA-GA/MR-GR - N) (INFERRED) (Disc)

Contrapositive: verse? – NOT in Mark, Luke, or John

1519. (if) you are persecute(d) and (have men to) say all manner of evil against you, for my sake, (then) Rejoice, and be exceeding glad: (since) . . . great is your reward in heaven . . . (since) . . . so persecuted they the prophets which were before you (MA-GA/MR-GR - N) (INFERRED) (Disc)

Contrapositive: verse? – NOT in Mark, Luke, or John

1520. (un)Till heaven and earth pass, (then) . . . one jot or one tittle shall in no wise pass from the law, (un)till all be fulfilled. (Disc) (GAGR) (NOT INFERRED)

Contrapositive: verse? – NOT in Mark, Luke, or John

1521. . . . when thou doest alms, (then) let not they left hand know what thy right hand doeth (Disc) (MAMR) (NOT INFERRED)

Contrapositive: verse? – NOT in Mark, Luke, or John

1522. . . . when thou doest alms, (then) let not they left hand know what thy right hand doeth (then) . . . That thine alms may be in secret: and thy Father which seeth in secret himself shall reward theee openly. (Disc) (MAMR) (NOT INFERRED)

Contrapositive: verse? – NOT in Mark, Luke, or John

1523. . . . when thou prayest, enter into thy closet, and when thous hast shut thy door, (then) pray to thy Father which is in secret; and thy Father which seeth in secret shall reward thee openly. (Disc) (MA/MR-GR) (INFERRED)

Contrapositive: verse? – NOT in Mark, Luke, or John

1524. . . . when ye pray, (then) use not vain repetitions as the heathen do, (since) . . . they think that shall be heard for their much speaking. (Disc) (MAMR) (INFERRED)

Contrapositive: verse? – NOT in Mark, Luke, or John

1525. . . . before ye ask him (your Father) . . . (,then) . . . your Father knoweth what things ye have need of . . . (Disc) (MAGR) (INFERRED)

Contrapositive: verse? – NOT in Mark, Luke, or John

1526. . . . (since) your Father knoweth what things ye have need of, before ye ask him, (then) . . . Be not ye therefore like unto them (the heathen who think they shall be head for their much speaking). (Disc) (MA/MR-GR) (INFERRED)

Contrapositive: verse? – NOT in Mark, Luke, or John

1527. (since you are not to pray to seen, to use vain repetitions, or to be unaware that God knows of your needs even before praying) . . . After this manner therefore pray ye: Our Father which art in heaven . . . For thine is the kingdom, the power, and the glory, forever, Amen. (Disc) (MAMR) (INFERRED)

Contrapositive: verse? – NOT in Mark or John

1528. (since) . . . Ye cannot serve God and mammon . . . (.then) . . . Therefore I say unto you, Take no thought for your life, what ye shall eat, or what ye shall drink: nor yet for your body, what ye shall put on. (Disc) (MAMR) (INFERRED)

Contrapositive: verse? – NOT in Mark, Luke, or John

1529. (since) . . . God shall clothe you . . . Therefore take no thought saying, What shall we eat? or What shall we drink?, or Wherewithal shall we be clothed? (Disc) (GAMR) (INFERRED)

Contrapositive: verseF? – NOT in Mark, Luke, or John

1530. . . . (since) . . . the morrow shall take thought for the things of itself, (then) . . . Take therefore no thought for the morrow. (Disc) (—MR) (INFERRED)

Contrapositive: verse? – NOT in Mark, Luke, or John

1531. . . . (since all these things shall be added unto you) . . . Take therefore no thought for the morrow . . . . . . (since) . . . the morrow shall take thought for the things of itself . . . (Disc) (GAMR) (INFERRED)

Contrapositive: verse? – NOT in Mark, Luke, or John

1532. If you judge (with a certain type of judgement), (then) . . . you shall be judged (by that same judgement) . . . (since) . . . with what judgment ye judge, ye shall be judged: and with what measure ye mete, it shall be measured to you again. (Disc) (I) (MAGR-F) (Inductive)

Contrapositive: verse? – NOT in Mark or John

1533. (since your Father which is in heaven give(s) good things, even to people who are evil) . . . Therefore all things whatsoever ye would that men do to you, do ye even so to them . . . (Disc) (GAMR) (INFERRED)

Contrapositive: verse? – NOT in Mark or John
1534. (since your Father which is in heaven give(s) good things, even to people who are evil) . . . Therefore all things whatsoever ye would that men do to you, do ye even so to them . . . (since) . . . this is the law and the prophets. (Disc) (GAMR) (INFERRED)

Contrapositive: verse? – NOT in Mark or John

1535. Enter ye in at the strait gate (since) wide is the gate, and broad is the way, that leadeth to destruction, and many there be that find it (and) Because strait is the gate, and narrow is the way, which leadeth to life, and few there be that find it. (Disc) (MA-GA/MR) (INFERRED)

Contrapositive: verse? – NOT in Mark, Luke, or John

{
1536. (if a thing is) . . . the wide gate (,then that thing is such that) . . . broad is the way that leadeth to destruction and many there be that find it. (Disc) (GA/GR-MR) (INFERRED)

Contrapositive: verse? – NOT in Mark, Luke, or John

1537. (if a thing is) . . . the strait gate (,then that thing is such that) . . . narrow is the way that leadeth to destruction and few there be that find it. (Disc) (GA/GR-MR) (INFERRED)

Contrapositive: verse? – NOT in Mark, Luke, or John

}close to STINV

1538. (since) . . . every good tree bringeth forth good fruit (only) and the same with corrupt trees (,then) . . . Wherefore by their fruits ye shall know them. (Disc) (MAMR) (INFERRED)

Contrapositive: verse? – NOT in Mark, Luke, or John

1539. (since those who say Lord, Lord, but doeth not my will shall not enter heaven) . . . Therefore whosever heareth these sayings of mine and doeth them, I will liken him unto a wise man, which built his house upon a rock . . . and every one that heareth these sayings and doeth them not, shall be likened unto a foolish man, which built his house upon the sand. (Disc) (MA-GA/MR-GR) (INFERRED)

Contrapositive: verse? – NOT in Mark or John

1540. (since) . . . (Jesus Christ saith . . . to the sick of the palsy,) Arise, take up they bed, and go unto thine house . . . (,then) . . . ye may know that the Son of man hath power on earth to forgive sins . . . (Relig) (GAMR) (INFERRED)

Contrapositive: verse? – NOT in Luke or John

1541. . . . go ye and learn what that (They that be whole need not a physician, but they that are sick) meaneth, I will have mercy, and not sacrifice . . . (since) . . . I am not come to call the righteous, but sinners to repentance. (Relig) (GAMR) (INFERRED)

Contrapositive: verse? – NOT in Mark, Luke, or John

1542. . . . Give place: (since) . . . the maid is not dead, but sleepeth. (Disc) (GAMR) (INFERRED)

Contrapositive: verse? – NOT in Mark, Luke, or John

1543. (since) . . . The harvest truly is plenteous, but the labourers are few; Pray ye therefore the Lord of the harvest, that he will send forth labourers into his harvest. (Disc) (GA-MA/MR-GR) (INFERRED)

Contrapositive: verse? – NOT in Mark, Luke, or John

1544. . . . as ye go (unto Israel only) (,then) preach, saying, The kingdom of heaven is at hand. Heal the sick, cleanse the lepers, raise the dead, cast out devils. (Disc) (MA/MR-GR) (INFERRED)

Contrapositive: verse? – NOT in Mark, Luke, or John

1545. . . . when ye come into an house, (then) . . . salute it. (Disc) (MAMR) (INFERRED)

Contrapositive: verse? – NOT in Mark, Luke, or John

1546. (since) . . . I (Jesus Christ) send you forth as sheep in the midst of wolves: be ye therefore wise as serpents, and harmless as doves.

Contrapositive: verse? – NOT in Mark, Luke, or John

1547. . . . beware of men (since) . . . they will deliver you up to the councils, and they will scourge you in their synagogues . . . (Disc) (MAMR) (INFERRED)

Contrapositive: verse? – NOT in Mark, Luke, or John

THE CONDITIONAL/PROCEDURAL FORMS IN THE BOOK OF ACTS

1548. (since) John truly baptized with water; but ye shall be baptized with the Holy Ghost not many days hence . . . (then) wait for the promise of the Father, which . . . ye have heard of me. (Disc) (MA-GA/MR) (INFERRED) (Neither inductive nor deductive)
Contrapositive: verse? – NOT in Matthew, Mark, Luke, or John

1549. . . . after that the Holy Ghost is come upon you . . . (then) ye shall receive power . . .
(Disc) (GAMR) (INFERRED) (Deductive)
Contrapositive: verse? – NOT in Matthew, Mark, Luke, or John

1550. . . . after that the Holy Ghost is come upon you . . . (then) ye shall receive power . . . (then) . . . ye shall be witnesses unto me in all Judea, and in Samaria, and unto the uttermost part of the earth.
(Disc) (GAMR) (INFERRED) (Deductive)
Contrapositive: verse? – NOT in Matthew, Mark, Luke, or John

1551. (since) .. he (Saul of Tarsus) prayeth, And hath seen in a vision a man named Ananias coming in, and putting his hand on him, that he might receive his sight . . . (then) Arise, and go into the street, which is called Straight, and enquire in the house of Judas for one called Saul of Tarsus (Indiv- Ananias) (MAMR) (INFERRED) (Neither inductive nor deductive)
Contrapositive: verse? – NOT in Matthew, Mark, Luke, or John

1552. (since) . . . he (Saul of Tarsus) is a chosen vessel unto me, to bear my name before the Gentiles, and kings, and the children of Israel . . . (then) Go thy (Ananias') way . . .
(Indiv- Ananias) (MA-GA/MR) (INFERRED) (Neither inductive nor deductive)
Contrapositive: verseF? – NOT in Matthew, Mark, Luke, or John

1553. (since) . . . I (Christ) will shew him how great things he must suffer for my name's sake . . . (then) (since) . . . he (Saul of Tarsus) is a chosen vessel unto me, to bear my name before the Gentiles, and kings, and the children of Israel . . . (then) Go thy (Ananias') way . . .
(Indiv- Ananias) (GA/GR-MR) (INFERRED) (Neither inductive nor deductive)
Contrapositive: verseF? – NOT in Matthew, Mark, Luke, or John

1554. (since) . . . I (Christ) am with thee, and no man shall set on thee to hurt thee . . . (then) Be not afraid, but speak, and hold not thy peace . . . (Indiv-Paul) (GAMR) (INFERRED) (Neither inductive nor deductive)
Contrapositive: verseF? – NOT in Matthew, Mark, Luke, or John

1555. . . . (since) I (Christ) have much people in this city . . . (then) (since) . . . I (Christ) am with thee, and no man shall set on thee to hurt thee . . . (then) Be not afraid, but speak, and hold not thy peace . . . (Indiv-Paul) (GA/GR-MR) (INFERRED) (Neither inductive nor deductive)
Contrapositive: verseF? – NOT in Matthew, Mark, Luke, or John

1556. (since) . . . they (the people of Jerusalem) will not receive thy (Paul's) testimony concerning me . . . (then) Make haste, and get thee (Paul) quickly out of Jerusalem . . . (Indiv-Paul) (MAMR) (INFERRED) (Neither inductive nor deductive)
Contrapositive: verseF? – NOT in Matthew, Mark, Luke, or John

1557. (since) . . . I (Christ) will send thee (Paul) far hence unto the Gentiles . . . (then) Depart (Paul) (Indiv-Paul) (GAMR) (INFERRED) (Neither inductive nor deductive)
Contrapositive: verseF? – NOT in Matthew, Mark, Luke, or John

1558. (since) . . . I (Christ) have appeared unto thee (Saul) for this purpose, to make thee (Saul) a minister and a witness both of these things which thou (Saul) hast seen, and of those things in the which I will appear unto thee (Saul); Delivering thee (Saul) from the people, and from the Gentiles, unto whom now I send thee (Saul), To open their eyes, and to turn them from darkness to light, and from the power of Satan unto God, that they may receive forgiveness of sins, and inheritance among them which are sanctified by faith that is in me (Christ) (Indiv-Saul) (GAGR) (INFERRED) (Neither inductive nor deductive)
Contrapositive: verse? – NOT in Matthew, Mark, Luke, or John

1559. (if through Saul's witness Christ) . . . open (s) their (the Gentiles) eyes, and . . . turn(s) them from darkness to light, and from the power of Satan unto God, (then) . . . they (the Gentiles) may receive forgiveness of sins, and inheritance among them which are sanctified by faith that is in me (Christ) (Indiv-Saul) (GAGR) (INFERRED) (Inductive)
Contrapositive: verseF? – NOT in Matthew, Mark, Luke, or John

THE CONDITIONAL FORMS IN THE BOOK OF I AND II CORINTHIANS

{
1560. (If a thing is) . . . This (bread from the Passover celebration with Christ and his disciples before Christ's Crucifixion) . . . , (then that thing is) . . . my (Christ's) body, which is broken for you . . . . (Believers) (GAGR) (INFERRED) (Neither inductive nor deductive)
Contrapositive: verseF? – NOT in Matthew, Mark, Luke, John, or Acts—see # 1445

(If a thing is) . . . my (Christ's) body . . . , (then that thing is) . . . This (bread from the Passover celebration with Christ and his disciples before Christ's Crucifixion) . . . . (Believers) (GAGR) (INFERRED) (Neither inductive nor deductive)
Contrapositive: verseF? – NOT in Matthew, Mark, Luke, John, or Acts – since this is a definition the converse also holds
}
{
1561. (If a thing is) . . . This cup . . . , (then that thing is) . . . the new testament in my blood . . . . (Believers) (GAGR) (INFERRED) (Neither inductive nor deductive)
Contrapositive: verseF? – NOT in Matthew, Mark, Luke, John, or Acts—see # 1446, which involves an exception to the formal rules of human reasoning.

(If a thing is) . . . the new testament in my blood . . . , (then that thing is) . . . This cup . . . . (Believers) (GAGR) (INFERRED) (Neither inductive nor deductive)
Contrapositive: verseF? – NOT in Matthew, Mark, Luke, John, or Acts – since this is a definition the converse also holds
}

1562. . . . as oft as ye (believers in Christ) drink it (This cup (which) is the new testament in my blood), . . . (then) this (drinking of "This cup") do ye

(believers in Christ) . . . in remembrance of me. (Believers) (MAMR) (INFERRED) (Neither inductive nor deductive)
Contrapositive: verseF? – NOT in Matthew, Mark, Luke, John, or Acts

1563. (since) . . . my (the Lord's) strength is made perfect in weakness . . . (then) My grace is sufficient for thee (Paul) . . . (Indiv-Paul) (GA-MA/GR-MR) (INFERRED) (Neither inductive nor deductive)
Contrapositive: verse? – NOT in Matthew, Mark, Luke, John, or Acts

THE CONDITIONAL FORMS IN THE BOOK OF THE REVELATION OF JOHN

{
1564. (If a human is Jesus Christ, then that thing is) . . . the first and the last . . . (Indiv-John) (GAGR) (INFERRED) (Neither inductive nor deductive)
Contrapositive: verseF? – NOT in Matthew, Mark, Luke, John, Acts, or I/II Corinthians – From statement attributed to Christ, ". . . I am the first and the last . . ."

(If a human is) . . . the first and the last . . . , (then that human is) . . . I (Jesus Christ). (Indiv-John) (GAGR) (INFERRED) (Neither inductive nor deductive)
Contrapositive: verseF? – NOT in Matthew, Mark, Luke, John, Acts, or I/II Corinthians – From statement attributed to Christ, ". . . I am the first and the last . . ." (which also results in converse holding since this is a definition
}
{
1565. (If a human is Jesus Christ, then that thing is) . . . he that liveth, and was dead . . . (Indiv-John) (GAGR) (INFERRED) (Neither inductive nor deductive)
Contrapositive: verseF? – NOT in Matthew, Mark, Luke, John, Acts, or I/II Corinthians – From statement attributed to Christ, ". . . I am he that liveth, and was dead . . ." {KEY NOTE: The issue of whether ". . . liveth, and was dead . . ." can be supported by science is a separate question from whether such a statement violates any formal rules of human reasoning. Specifically, no formal rules of human reasoning are violated by such a statement, and there are varying opinions from different parts of the scientific community regarding the truth of personal testimonies with respect to reported 'life-after-death' or 'near-death' experiences.

(If a human is) . . . he that liveth, and was dead . . . , (then that human is) . . . I (Jesus Christ). (Indiv-John) (GAGR) (INFERRED) (Neither inductive nor deductive)
Contrapositive: verseF? – NOT in Matthew, Mark, Luke, John, Acts, or I/II Corinthians – From statement attributed to Christ, ". . . I am he that liveth, and was dead . . ." (which also results in converse holding since this is a definition)
}
{
1566. (If things are) . . . the seven stars . . . , (then those things) . . . are the angels of the seven churches . . . . (Indiv-John) (GAGR) (INFERRED) (Deductive)
Contrapositive: verseF? – NOT in Matthew, Mark, Luke, John, Acts, or I/II Corinthians

(If those things) . . . are the angels of the seven churches . . . , (then those things are) . . . the seven stars . . . . (Indiv-John) (GAGR) (INFERRED) (Deductive)
Contrapositive: verseF? – NOT in Matthew, Mark, Luke, John, Acts, or I/II Corinthians – since this is a definition the converse also holds
}
{
1567. (If things are) . . . the seven candlesticks . . . , (then those things) . . . are the seven churches . . . . (Indiv-John) (GAGR) (INFERRED) (Deductive)
Contrapositive: verseF? – NOT in Matthew, Mark, Luke, John, Acts, or I/II Corinthians

(If those things) . . . are the seven churches . . . , (then those things are) . . . the seven . . . . (Indiv-John) (GAGR) (INFERRED) (Deductive)
Contrapositive: verseF? – NOT in Matthew, Mark, Luke, John, Acts, or I/II Corinthians – since this is a definition the converse also holds
}

1568. . . . (since) thou (the angel of the church of Ephesus) hast left thy first love . . . (then) I (Christ) have somewhat against thee (the angel of the church of Ephesus) . . . (Indiv-John) (MAGR) (INFERRED) (Neither inductive nor deductive)
Contrapositive: verseF? – NOT in Matthew, Mark, Luke, John, Acts, or I/II Corinthians

1569. . . . (since) thou (the angel of the church of Ephesus) hast left thy first love . . . (then) I (Christ) have somewhat against thee (the angel of the church

of Ephesus) . . . therefore from whence thou art fallen, and repent, and do the first works . . . (Indiv-John) (MAGR) (INFERRED) (Neither inductive nor deductive)
Contrapositive: verseF? – NOT in Matthew, Mark, Luke, John, Acts, or I/II Corinthians

1570. (if thou do not) . . . Remember . . . from whence thou art fallen, and repent, and do the first works . . . (then) else I will come unto thee (the angel of the church of Ephesus) quickly, and will remove thy candlestick out of his place(Indiv-John) (MAGR) (INFERRED) (Neither inductive nor deductive)
Contrapositive: verseF? – NOT in Matthew, Mark, Luke, John, Acts, or I/II Corinthians

1571. (if) thou . . . (do not) repent (then) . . . else I will come unto thee (the angel of the church of Ephesus) quickly, and will remove thy candlestick out of his place . . . (Indiv-John) (MAGR) (INFERRED) (Neither inductive nor deductive)
Contrapositive: verseF? – NOT in Matthew, Mark, Luke, John, Acts, or I/II Corinthians

1572. (if one) . . . hath an ear, (then) . . . let him hear what the Spirit saith unto the churches (Indiv-John) (GAMR) (INFERRED) (Deductive)
Contrapositive: verseF? – NOT in Matthew, Mark, Luke, John, Acts, or I/II Corinthians

1573. (if one) . . . overcometh . . . (then) . . . will I (Christ) give . . . To him . . . to eat of the tree of life . . . (Indiv-John) (MAGR) (INFERRED) (Deductive)
Contrapositive: verseF? – NOT in Matthew, Mark, Luke, John, Acts, or I/II Corinthians

1574. (if) thou (are) faithful unto death, (then) I will give thee (the angel of the church in Smyrna) a crown of life. . . . (Indiv-John) (MAGR) (INFERRED) (Neither inductive nor deductive)
Contrapositive: verseF? – NOT in Matthew, Mark, Luke, John, Acts, or I/II Corinthians

1575. (if one) . . . overcometh . . . (then that one) . . . shall not be hurt of the second death (Indiv-John) (MAGR) (INFERRED) (Deductive)
Contrapositive: verseF? – NOT in Matthew, Mark, Luke, John, Acts, or I/II Corinthians

1576. . . . (since) thou (the angel of the church in Pergamos) hast there them that hold the doctrine of Balaam . . . (and since) hast thou also them that hold the doctrine of the Nicolaitanes . . . (then) I (Christ) have a few things against thee . . . (Indiv-John) (MAGR) (INFERRED) (Neither inductive nor deductive)
Contrapositive: verseF? – NOT in Matthew, Mark, Luke, John, Acts, or I/II Corinthians

1577. (if) thou . . . (do not) repent (then) . . . else I will come unto thee quickly, fight against them (those who hold the doctrines of Balaam and Nicolaitanes) with the sword of my mouth . . . (Indiv-John) (MAGR) (INFERRED) (Neither inductive nor deductive)
Contrapositive: verseF? – NOT in Matthew, Mark, Luke, John, Acts, or I/II Corinthians

1578. (if one) . . . overcometh . . . (then that one) . . . will I (Christ) give to eat of the hidden manna, and will give him a white stone, and in the stone a new name written, which no man knoweth saving him that receiveth it. (Indiv-John) (MAGR) (INFERRED) (Deductive)
Contrapositive: verseF? – NOT in Matthew, Mark, Luke, John, Acts, or I/II Corinthians

1579. . . . (since) thou (the angel of the church in Thyatira) sufferest that woman Jezebel . . . (then) I (Christ) have a few things against thee . . . (Indiv-John) (MAGR) (INFERRED) (Neither inductive nor deductive)
Contrapositive: verseF? – NOT in Matthew, Mark, Luke, John, Acts, or I/II Corinthians

1580. (if) they (who sufferest that woman Jezebel) . . . (do not) repent (then) . . . else I will cast her into a bed, and them that commit adultery with her into great tribulation . . . (Indiv-John) (MAGR) (INFERRED) (Neither inductive nor deductive)
Contrapositive: verseF? – NOT in Matthew, Mark, Luke, John, Acts, or I/II Corinthians—see # 1583 for close to Inverse

1581. (when) I will cast her into a bed, and them that commit adultery with her into great tribulation . . . (then) all the churches shall know that I am he which searcheth the reins and hearts . . . (Indiv-John) (MAGR) (INFERRED) (Neither inductive nor deductive)
Contrapositive: verseF?—NOT in Matthew, Mark, Luke, John, Acts, or I/II Corinthians

{
1582. (If a human is Jesus Christ, then that thing is) . . . he which searcheth the reins and hearts . . . (Indiv-John) (GAGR) (INFERRED) (Neither inductive nor deductive)
Contrapositive: verseF? – NOT in Matthew, Mark, Luke, John, Acts, or I/II Corinthians – From statement attributed to Christ, ". . . I am he which searcheth the reins and hearts . . ."

(If a human is) . . . he which searcheth the reins and hearts . . . , (then that human is) . . . I (Jesus Christ). (Indiv-John) (GAGR) (INFERRED) (Neither inductive nor deductive)
Contrapositive: verseF? – NOT in Matthew, Mark, Luke, John, Acts, or I/II Corinthians – From statement attributed to Christ, ". . . I am he which searcheth the reins and hearts . . ." (which also results in converse holding since this is a definition)
}

1583. close to Inverse of # 1580 just above – Contrapositive: verse?
STINV

1584. (if) . . . ye have already (a thing) . . . (then) hold fast (that thing) till I come. (Indiv-John) (MA-GA/MR) (INFERRED) (Neither inductive nor deductive)
Contrapositive: verseF? – NOT in Matthew, Mark, Luke, John, Acts, or I/II Corinthians

1585. till I come (,then) hold fast (that thing). (Indiv-John) (GAMR) (INFERRED) (Neither inductive nor Deductive)
Contrapositive: verseF? – NOT in Matthew, Mark, Luke, John, Acts, or I/II Corinthians

1586. (if one) . . . overcometh and keepeth my works unto the end . . . (then) . . . to him will I (Christ) give power over the nations: And He shall rule them with a rod of iron . . . And I (Christ) will give him the morning star. (Indiv-John) (MA/GR-MR) (INFERRED) (Deductive)
Contrapositive: verse? – NOT in Matthew, Mark, Luke, John, Acts, or I/II Corinthians

1587. (since) . . . I (Christ) have not found thy (the angel of the church at Sardis) works perfect before God . . . (then) Be watchful, and strengthen the things which remain, that are ready to die . . . (Indiv-John) (GAMR) (INFERRED) (Neither inductive nor deductive)

Contrapositive: verseF? – NOT in Matthew, Mark, Luke, John, Acts, or I/II Corinthians

1588. (since) . . . I (Christ) have not found they works perfect before God . . . (then) Be watchful, and strengthen the things which remain, that are ready to die . . . Remember therefore how thou (the angel of the church in Sardis) hast received and heard, and hold fast, and repent. (Indiv-John) (GAMR) (INFERRED) (Neither inductive nor deductive)
Contrapositive: verseF? – NOT in Matthew, Mark, Luke, John, Acts, or I/II Corinthians

1589. (since) . . . I (Christ) have not found they works perfect before God . . . (then) Be watchful, and strengthen the things which remain, that are ready to die . . . Remember therefore how thou (the angel of the church in Sardis) hast received and heard, and hold fast, and repent If therefore thou shalt not watch, (then) I (Christ) will come on thee as a thief, and thou shalt not know what hour I will come upon thee. (Indiv-John) (GA-MA/MR-GR) (INFERRED) (Neither inductive nor deductive)
Contrapositive: verseF? – NOT in Matthew, Mark, Luke, John, Acts, or I/II Corinthians

1590. If therefore thou (the angel of the church at Sardis) shalt not watch, (then) I (Christ) will come on thee as a thief, and thou shalt not know what hour I will come upon thee. (Indiv-John) (MA/GR-MR) (INFERRED) (Neither inductive nor deductive)
Contrapositive: verse? – NOT in Matthew, Mark, Luke, John, Acts, or I/II Corinthians

1591. (if one has) . . . not defiled (one's) garments (,then that one) shall walk with me in white . . . (Indiv-John) (MAMR) (INFERRED) (Deductive)
Contrapositive: verse? – NOT in Matthew, Mark, Luke, John, Acts, or I/II Corinthians

1592. (since) they (which have not defiled their garments) are worthy . . . (then) they shall walk with me in white . . . (Indiv-John) (MAMR) (INFERRED) (Inductive)
Contrapositive: verse? – NOT in Matthew, Mark, Luke, John, Acts, or I/II Corinthians

1593. (if one) . . . overcometh . . . (then that one) . . . shall be clothed in white raiment; and I will not blot out his name out of the book of life, but I will con-

fess his name before my Father, and before his angels. (Indiv-John) (MAGR) (INFERRED) (Deductive)
Contrapositive: verse? – NOT in Matthew, Mark, Luke, John, Acts, or I/II Corinthians

{
1594. (if) he (Christ) . . . openeth (a thing) . . . (,then) no man shutteth (that thing) . . . (Indiv-John) (MAGR) (INFERRED) (Neither inductive nor deductive)
Contrapositive: verseF? – NOT in Matthew, Mark, Luke, John, Acts, or I/II Corinthians

1595. (if) he (Christ) . . . shutteth (a thing) . . . (,then) no man openeth (that thing) . . . (Indiv-John) (MAGR) (INFERRED) (Neither inductive nor deductive)
Contrapositive: verseF? – NOT in Matthew, Mark, Luke, John, Acts, or I/II Corinthians
STINV

1596. (since) . . . thou (the angel of the church in Philadelphia) has a little strength, and hast kept my word, and hast not denied my name . . . (then) behold I (Christ) have set before thee an open door, and no man can shut it . . . (Indiv-John) (MA/GR-MR) (INFERRED) (Neither inductive nor deductive)
Contrapositive: verse? – NOT in Matthew, Mark, Luke, John, Acts, or I/II Corinthians

1597. (since) thou (the angel of the church in Philadelphia) hast kept the word of my patience, (then) I (Christ) also will keep thee from the hour of temptation, which shall come upon all of the world, to try them that dwell upon the earth. (Indiv-John) (MAGR) (INFERRED) (Neither inductive nor deductive)
Contrapositive: verse? – NOT in Matthew, Mark, Luke, John, Acts, or I/II Corinthians

1598. (if one) . . . hold fast which (one) hast (then) . . . no man take thy crown. (Indiv-John) (MAMR) (INFERRED) (Deductive)
Contrapositive: verse? – NOT in Matthew, Mark, Luke, John, Acts, or I/II Corinthians

1599. (if one) . . . overcometh . . . (then that one) . . . will I (Christ) make a pillar in the temple of my God, and he shall go no more out: and I will write

upon him the name of my God, and the name of the city of my God, which is new Jerusalem, which cometh down out of heaven from my God: and I will write upon him my new name. (Indiv-John) (MAGR) (INFERRED) (Deductive)
Contrapositive: verse? – NOT in Matthew, Mark, Luke, John, Acts, or I/II Corinthians

{
1600. (if a thing is) . . . the name of the city of my God . . . (, then that thing is) .. new Jerusalem. (Indiv-John) (GAGR) (INFERRED) (Neither inductive nor deductive)
Contrapositive: verse? – NOT in Matthew, Mark, Luke, John, Acts, or I/II Corinthians - from the statement: ". . . the name of the city of my God, which is new Jerusalem . . ."

(if a thing is) . . . the name of the city of my God . . . (, then that thing is) .. new Jerusalem. (Indiv-John) (GAGR) (INFERRED) (Neither inductive nor deductive)
Contrapositive: verse? – NOT in Matthew, Mark, Luke, John, Acts, or I/II Corinthians – since this is a definition the converse also holds from the statement: ". . . the name of the city of my God, which is new Jerusalem . . ."
}

1601. (since) . . . thou (the angel of the church in Laodicea) art lukewarm, and neither cold nor hot, (then) . . . I (Christ) will spue thee out of my mouth . . . (Indiv-John) (MAGR) (INFERRED) (Neither inductive nor deductive)
Contrapositive: verse? – NOT in Matthew, Mark, Luke, John, Acts, or I/II Corinthians

1602. (since) . . . thou (the angel of the church in Laodicea) sayest, I am rich, and increased with goods, and have need of nothing; and knowest not that thou art wretched, and miserable, and poor, and blind, and naked . . . (then) I counsel thee to buy of me (Christ) gold tried in the fire, that thou mayest be rich; and white raiment, that thou mayest be clothed, and that the shame of thy nakedness do not appear; and anoint thine eyes with eyesalve, that thou mayest see. (Indiv-John) (MA/MR-GR) (INFERRED) (Neither inductive nor deductive)
Contrapositive: verseF? – NOT in Matthew, Mark, Luke, John, Acts, or I/II Corinthians

1603. (if one) . . . buy of me (Christ) gold tried in the fire, (then) . . . thou mayest be rich . . . (Indiv-John) (MA-GA/MR) (INFERRED) (Deductive)

Contrapositive: verseF? – NOT in Matthew, Mark, Luke, John, Acts, or I/II Corinthians

1604. (if one) . . . anoint (one's) eyes with eyesalve . . . , (then) . . . thou mayest see . . . (Indiv-John) (MAMR) (INFERRED) (Deductive)
Contrapositive: verseF? – NOT in Matthew, Mark, Luke, John, Acts, or I/II Corinthians

1605. (if one) . . . buy of me (Christ) gold tried in , (then) . . . thou mayest be rich . . . (Indiv-John) (MA-GA/MR) (INFERRED) (Deductive)
Contrapositive: verseF? – NOT in Matthew, Mark, Luke, John, Acts, or I/II Corinthians

1606. (if) .. I (Christ) love (one) . . . (,then that one will Christ) rebuke and chasten . . . (Indiv-John) (GAGR) (INFERRED) (Neither inductive nor deductive)
Contrapositive: verse? – NOT in Matthew, Mark, Luke, John, Acts, or I/II Corinthians

1607. (if) .. I (Christ) love (one) . . . (,then that one will Christ) rebuke and chasten . . . therefore . . . repent. (Indiv-John) (GAGR) (INFERRED) (Neither inductive nor deductive)
Contrapositive: verseF? – NOT in Matthew, Mark, Luke, John, Acts, or I/II Corinthians

1608. . . . if any man hear my voice, and open the door, (then) . . . I (Christ) will come in to him, and will sup with him, and he with me. (Indiv-John) (MA-GA/GR) (INFERRED) (Deductive)
Contrapositive: verseF? – NOT in Matthew, Mark, Luke, John, Acts, or I/II Corinthians

1609. (if one) . . . overcometh . . . (then that one) . . . will I (Christ) grant to sit with me in my throne, even as I overcame, and am set down with my Father in his throne. (Indiv-John) (MAGR) (INFERRED) (Deductive)
Contrapositive: verse? – NOT in Matthew, Mark, Luke, John, Acts, or I/II Corinthians

{
1610. (If a human is Jesus Christ, then that thing is) . . . Alpha and Omega, the beginning and the end, the first and the last . . . (Indiv-John) (GAGR) (INFERRED) (Neither inductive nor deductive)

Contrapositive: verseF? – NOT in Matthew, Mark, Luke, John, Acts, or I/II Corinthians – From statement attributed to Christ, ". . . I am Alpha and Omega, the beginning and the end, the first and the last . . ."

(If a human is) . . . Alpha and Omega, the beginning and the end, the first and the last . . . , (then that human is) . . . I (Jesus Christ). (Indiv-John) (GAGR) (INFERRED) (Neither inductive nor deductive)
Contrapositive: verseF? – NOT in Matthew, Mark, Luke, John, Acts, or I/II Corinthians – From statement attributed to Christ, ". . . I am Alpha and Omega, the beginning and the end, the first and the last . . ." (which also results in converse holding since this is a definition
}
{
1611. (If a human is Jesus Christ, then that thing is) . . . the root and the offspring of David and the bright and morning star . . . (Indiv-John) (GAGR) (INFERRED) (Neither inductive nor deductive)
Contrapositive: verseF? – NOT in Matthew, Mark, Luke, John, Acts, or I/II Corinthians – From statement attributed to Christ, ". . . I am the root and the offspring of David and the bright and morning star . . ."

(If a human is) . . . the root and the offspring of David and the bright and morning star . . . , (then that human is) . . . I (Jesus Christ). (Indiv-John) (GAGR) (INFERRED) (Neither inductive nor deductive)
Contrapositive: verseF? – NOT in Matthew, Mark, Luke, John, Acts, or I/II Corinthians – From statement attributed to Christ, ". . . I am the root and the offspring of David and the bright and morning star . . ." (which also results in converse holding since this is a definition
}

ADDITIONAL STATEMENTS OF INCLUSION (ONE-WAY AND DUAL)

1612. A STATEMENT OF INCLUSION IN ONE DIRECTION: "In Him (the Word or Christ) was life . . ." – John 1: 4 - (Relig) (GA__) (INFERRED) (Neither inductive nor deductive)
Contrapositive: verse?

1613. A STATEMENT OF INCLUSION IN ONE DIRECTION: ". . . the water that I (Christ) shall give him (who drinketh of the water that Christ giveth) shall be in him (who drinketh of the water . . .) a well of water springing up

into everlasting life." – John 4: 23 - (Relig) (GA__) (INFERRED) (Neither inductive nor deductive)
Contrapositive: verse?

1614. A STATEMENT OF INCLUSION IN ONE DIRECTION: ". . . no unrighteousness is in him (he that seeketh his glory that sent him) – John 7: 18 - (Relig) (__GR) (INFERRED) (Neither inductive nor deductive)
Contrapositive: verse?

1615. A STATEMENT OF INCLUSION IN ONE DIRECTION: ". . . there is no truth in him (the devil)." – John 8: 44 - (Relig) (____) (INFERRED) (Neither inductive nor deductive)
Contrapositive: verse?

1616. A STATEMENT OF DUAL INCLUSION: ". . . the Father is in me . . ." and ". . . I (Christ) in him)." So that (if one is) . . . (God) the Father". . . (,then that one) . . . is in me (Christ) and the converse– John 10: 38 (Relig) (GAGR) (INFERRED) (Neither inductive nor deductive)
Contrapositive: verse?

1617. A STATEMENT OF INCLUSION IN ONE DIRECTION: ". . . there is no light in him (who walk(s) in the night)." – John 11: 10 - (Relig) (__GR) (INFERRED) (Neither inductive nor deductive)
Contrapositive: verse?

1618. A STATEMENT OF INCLUSION IN ONE DIRECTION: ". . . God (the Father) is glorified in him (Christ)." – John 13: 31 - (Relig) (GAGR) (INFERRED) (Neither inductive nor deductive)
Contrapositive: verse?

1619. A STATEMENT OF DUAL INCLUSION: ". . . I (Christ) in him (the one who abideth in Christ) . . ."; along with "He that abideth in me (Christ) . . ." so that (if one is) . . . I (Christ) . . . (,then that one is) . . . in him (a disciple who abideth in Christ) and the converse– John 15: 5 (Relig) (GAMR & MAGR) (INFERRED) (Neither inductive nor deductive)
Contrapositive: verse?

1620. A STATEMENT OF DUAL INCLUSION: ". . . I (Christ) am in the Father . . ." and ". . . the Father in me (Christ)." So that (if one is) . . . I (Christ) . . . (,then that one is) . . . in (God) the Father) . . . and the converse – John 14: 10 (Relig) (GAGR) (INFERRED) (Neither inductive nor deductive)
Contrapositive: verse?

1621. A STATEMENT OF INCLUSION IN ONE DIRECTION: ". . . I (Christ) am in my Father." – John 14: 20 - (Relig) (GAGR) (INFERRED) (Neither inductive nor deductive)
Contrapositive: verse?

1622. A STATEMENT OF DUAL INCLUSION: ". . . ye (the 11 Apostles) in me (Christ)."; along with ". . . I (Christ) in you (the 11 Apostles)." So that (if one is a member of) . . . ye (the 11 Apostles) . . . (,then that one is) . . . in me (Christ) . . . and the converse– John 14: 20 - (Relig) (MAGR & GAMR) (INFERRED) (Neither inductive nor deductive)
Contrapositive: verse?

1623. A STATEMENT OF INCLUSION IN ONE DIRECTION: ". . . in me (Christ) ye (disciples of Christ) might have peace . . ."– John 16: 33 - (Relig) (GA__) (INFERRED) (Neither inductive nor deductive)
Contrapositive: verse?

1624. A STATEMENT OF DUAL INCLUSION: ". . . thou Father art in me (Christ) . . ." and ". . . I (Christ) in thee." (if one is) . . . thou (God the) Father". . . (,then that one is) . . . in me (Christ) and the converse – John 17: 21 - (Relig) (GAGR) (INFERRED) (Neither inductive nor deductive)
Contrapositive: verse?

1625. A STATEMENT OF INCLUSION IN ONE DIRECTION: ". . . I (Christ) in them (disciples of Christ)." – John 17: 23 - (Relig) (GAMR) (INFERRED) (Neither inductive nor deductive)
Contrapositive: verse?

1626. A STATEMENT OF INCLUSION IN ONE DIRECTION: ". . . thou (God the Father) in me." – John 17: 23 - (Relig) (GAGR) (INFERRED) (Neither inductive nor deductive)
Contrapositive: verse?

NOTE FOR "ADDITIONAL CONDITIONAL, PROCEDURAL STATEMENTS" – See # 1495 and # 1496 for the two (2) other statements related to inclusion

ADDITIONAL CONDITIONAL FORMS NOTED IN LUKE:

1627. (If a thing exists) . . . , (then that thing) . . . shall (not) by any means harm you (the disciples) . . . . (Disciples) (GAMR) (INFERRED) (Deductive)
The converse holds since this is a definition.

Contrapositive: verseF? – ONLY in Luke – See Luke 10: 19

1628. (when) . . . they shall lay their hands on you, and persecute you, delivering you up to the synagogues, and into prisons, being brought before kings and rulers for my name's sake . . . And ye shall be betrayed both by parents, and brethren, and kinsfolks, and friends; and some of you shall they cause to be put to death. And ye shall be hated of all men for my name's sake (,then) . . . shall not an hair of your (the disciples) head perish." (Disciples) (MA/GR-MR) (INFERRED) (Neither inductive nor Deductive)

Contrapositive: verseF? - ONLY in Luke – See Luke 21: 12–18

1629. . . . before all these things (during which Nation shall rise against nation, and kingdom against kingdom: And great earthquakes shall be in divers places, and famines, and pestilences; and fearful sights and great signs shall there be from heaven, (then) they (those who oppose Christ) shall lay their hands on you, and persecute you, delivering you up to the synagogues, and into prisons, being brought before kings and rulers for my name's sake.And it shall turn to you for a testimony . . . And ye shall be betrayed both by parents, and brethren, and kinsfolks, and friends; and some of you shall they cause to be put to death. And ye shall be hated of all men for my name's sake . . . (Disciples) (MA-GA/MR) (INFERRED) (Neither inductive nor Deductive)

Contrapositive: verseF? - ONLY in Luke – See Luke 21: 10–16

www.ingramcontent.com/pod-product-compliance
Ingram Content Group UK Ltd.
Pitfield, Milton Keynes, MK11 3LW, UK
UKHW041103150926
475561UK00001BA/21